Intranet Performance
MANAGEMENT

The CRC Press
Advanced and Emerging Communications
Technologies Series

Series Editor-in-Chief: Saba Zamir

Intranet Performance
MANAGEMENT

Kornel Terplan

Series Editor-in-Chief
Saba Zamir

CRC Press
Boca Raton London New York Washington, D.C.

Library of Congress Cataloging-in-Publication Data

Terplan, Kornel.
    Intranet performance management / Kornel Terplan.
        p. cm. — (The CRC Press advanced and emerging communications technologies series)
     ISBN 0-8493-9200-4 (alk. paper)
     1. Intranets (Computer networks)—Management. I. Title.
  II. Series.
  TK5105.875.I6T46 1999
  004.6'8—dc21
                                                 99-38584
                                                   CIP

No claim to original U.S. Government works
International Standard Book Number 0-8493-9200-4
Library of Congress Card Number 99-38584
Printed in the United States of America 1 2 3 4 5 6 7 8 9 0
Printed on acid-free paper

Preface

The use of Internet technology is changing business practices in each industry segment. The opportunities offered by electronic commerce are not yet fully used in today's implementations. Working styles and productivity of employees are significantly changing due to centralization and simplification of access to internal company documentation. In addition, real-time inquiry options can be supported by this technology. Information distribution capabilities to external users, sometimes called visitors, are changing as well. In both cases, the quality of page content, navigation between pages, page design, and page maintenance are critical success factors.

Network managers and system administrators welcome the Internet technology, but want to implement it in a more secure environment. The result of this "privatization" process is intranets and extranets. This book concentrates on intranet performance management. Other management functions, such as fault, configuration, security, and accounting management, are addressed as well, but without details. In managing intranets, many existing tools and emerged management technologies may be used and reimplemented. The most significant challenges have been observed with performance and security management.

Intranet management functions are summarized in Chapter 1. It highlights not only the functions of fault, performance, configuration, security, and accounting management, but also the expected challenges due to Internet technology. The most significant changes are expected in the areas of performance and security management. User behavior and workload patterns change in intranet environments, resulting in uncertainties in service and usage predictions. The intelligent use of firewalls requires very accurate risk analysis and protection extensions in intranets and extranets. Many known procedures in fault, configuration, and accounting management can be reimplemented in intranets. This book focuses on performance management.

The basics of Internet technology are briefly outlined in Chapter 2. This short introduction incorporates Web servers, Web browsers, their communication protocol (HTTP), and their content presentation language (HTML). It also covers the basics of the underlying protocol architecture (TCP/IP), in particular, its opportunities and limitations. The closing part of this chapter focuses on future extensions, such as DHTML, XML, Perl, TCL, and HTML for voice and video.

Critical success factors for intranet management are discussed in Chapter 3. Each component of intranets, such as Web servers, Web browsers, access networks, backbone networks, and interconnecting devices, is investigated in depth, whether it causes performance bottlenecks, and if yes, how to avoid them. The first experiences with managing intranets indicate that content-driven link management, load balancing, resource management, and flow admission control are extremely important.

Chapter 4 is devoted completely to content management. All phases of content management, such as content authoring, publishing, searching, distributing, and analyzing content, are addressed in depth. In particular, site and page design considerations are discussed and practical recommendations given. In addition, a survey on content authoring and deployment tools is included to help Webmasters and Web administrators evaluate and select the right mix of tools. It is important to highlight opportunities, to integrate these tools with usage analysis tools and management platforms.

Chapter 5 explores in detail the use of log file analysis to generate navigation patterns, contents requested, visitor information, and fulfillment analysis for Web sites. Following these basic goals, the steps of usage analysis are displayed, followed by critical issues and limitations of log file analysis. The industry offers a rich set of products, but with different functionality. A survey of log analysis tools is included to help Webmasters and Web administrators evaluate and select the right mix of tools. Some of the tools offer integration with content authoring and deployment tools; others prefer integration with load balancing solutions. Integration with management platforms also is referenced.

Traffic monitors working on the digital interface of communication infrastructures are evaluated in Chapter 6. They are independent from underlying hardware, software, and communications protocols, and offer low-overhead data capturing. Network managers find a limited selection of tools on the market. A survey of traffic monitors is included to help Webmasters and Web administrators evaluate and select the right tools. Some of the tools offer integration capabilities with software-based solutions. Thus, Webmasters and Web administrators may obtain a more complete performance review of intranet behavior.

Generic Web server management does not differ significantly from managing other kinds of servers. Chapter 7 deals with the management of Unix- and NT-based Web servers. Leading products for both areas are evaluated using a rich list of criteria. Particular attention is given to platform tools, such as OpenView from Hewlett Packard and Unicenter TNG from Computer Associates, which can manage both Unix and NT environments. Emerging technologies to distribute Web server load and to optimize access performance also are addressed. Content driven replication and distribution will significantly improve overall performance of Web server farms. The first product examples are included in the closing section of this chapter.

Load balancing and traffic shaping are the key topics addressed in Chapter 8. Using Internet technology, traffic patterns, user behavior, and careful control of bandwidth in backbone and access networks are the critical factors. There is a definite need for tighter bandwidth control in order to meet service level requirements. Load balancing helps to utilize intranet resources more effectively. Products in this category are hardware or software based, or a combination of both. A survey of load balancing and traffic shaping tools is included to help evaluate and select the right mix of tools. Some of the tools are stand-alone; others may be embedded into other networking equipment, such as routers, switches, and firewalls. Most products are easily manageable via SNMP agents.

Service level agreements use response time and availability as principal metrics. Chapter 9 focuses on measuring these indicators from the end user perspective.

Response time definition is followed by the four basic response time measurement alternatives: use of monitors and analyzers, use of synthetic workload, use of application agents, and use of application response measurement (ARM) MIBs. A couple of emerging products are evaluated in depth. Common to all of them is that service metrics are related to business applications. Not just individual components, but all components are included in the response time path between users and applications residing on Web servers. Most of the products referenced here may be used in combination with traffic monitors, load balancers, and traffic shapers. The trend is that these tools will be integrated with management platforms and modeling packages.

The final part of the book, Chapter 10, predicts trends in managing intranets. In addition to assessing business changes and organizational changes for defining new responsibilities for Internet technologies, integration needs and opportunities are outlined. Intranet management is considered as a new mixture of emerged and emerging technologies. Many existing techniques can be redeployed and expanded by completely new products focused on intranets and their performance.

Acknowledgments

I would like to thank Chuck Williams, Vice President; Susan Greger, Director; and Robert Renner, Director, from Georgia-Pacific Corporation for brainstorming sessions; Endre Sara from Stevens Institute of Technology for pre-evaluating management products; and Adam Szabo for preparing the artwork for the book.

Three vendors were extremely helpful, supporting me with up-to-date product information: Suzanne Baylor from WebTrends, Steve Ballerini from Andromedia, and John Meier from Freshtech.

Special thanks are due to Dawn Mesa, associate editor; Bill Heyward, project editor; and Suzanne Lassandro of CRC Press, who were extremely helpful in every phase of this publication.

Kornel Terplan

Table of Contents

Chapter 5

Chapter 6

1 Introduction

CONTENTS

1.1 INTRODUCTION

Intranet management means deploying and coordinating resources to design, plan, administer, analyze, operate, and expand intranets to meet service level objectives at all times, at reasonable cost, and with optimal capacity of resources. Intranet managers can use all the experiences collected over the last 25 years managing data networks. Existing management concepts are still valid. Critical success factors are applicable as well. In managing intranets, those critical success factors include (Terplan, 1996):

- Management processes that can be grouped around fault, configuration, performance, security, and accounting management
- Management tools that are responsible for supporting management processes and are usually assigned to human resources
- Human resources of the management team, with their skills and network management experiences

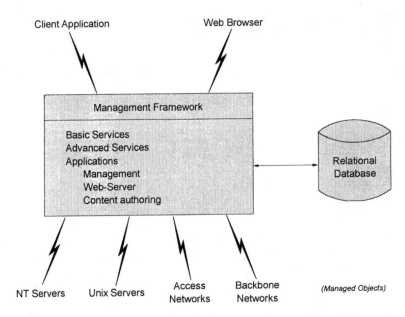

FIGURE 1.1   Intranet management framework.

Intranet management instrumentation shows similarities with the management of other networks. The architecture is shown in Figure 1.1 (Terplan, 1998b). The management framework is the center and is responsible for consolidating, processing, displaying, and distributing information to authorized persons. The framework is expected to be equipped with Web capabilities meeting the expectations of the majority of users. This means that views and reports are converted into HTML (hypertext markup language) pages and are accessible from universal browsers. Management applications are a mix of well-known ones, such as trouble ticketing, asset management, and change management, and brand new ones, dealing with log file analysis, load balancing, traffic shaping, content authoring, and Web server management.

The remainder of this chapter addresses specific challenges of intranet management toward management processes.

1.2   INTERNET, INTRANETS, AND EXTRANETS

The Internet is an existing network used by millions of people every day. In addition, the "Internet" is a generic term for a bundle of technologies available under the Internet umbrella. The Internet is very similar to the global phone system. In the phone network, whoever is a subscriber can be reached by dialing the right country code, area code, and local phone number. In the case of the Internet, visitors type in the right universal resource locator (URL) to access the necessary information. Even the billing process shows similarities; the longer the person talks or surfs, the higher the bill.

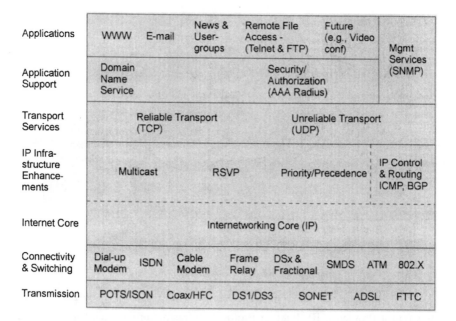

FIGURE 1.2 Typical layers of the communication architecture.

Ownership is not so clear with the Internet as with public phone systems. There are multiple owners of the Internet physical backbone; but they are hidden from users. Administration and management are increasingly important because the number of subscribers is growing very fast. Just one administration issue — address management — causes a lot of headaches. Country institutions responsible for distributing IP addresses locally are coordinated by an independent U.S.-based company. Basically, the Internet can support multiple communication forms, such as voice, data, and video. The predominant use is still data.

Internet users today use a small but powerful set of applications. The most widely used applications are e-mail, World Wide Web (WWW) content browsing, and file transfer services. These services are provided by high-powered servers within the network and the software implementing these applications (e.g., Web page hosting and mail forwarding). Figure 1.2 shows the typical layers of the communication architecture.

Application support services — such as Domain Name Service, which converts the widely seen host names such as "www.company.com" into IP addresses — and authorization services provide support across applications. There are also software applications running on network servers. Transport services provide the option for reliable transport of information (error detection and retransmissions) or simple unacknowledged transfer. These services may operate end-to-end (e.g., in case of file transfers between and end user and a remote Internet server). In such cases, the network doesn't get involved at the transport level. In other cases (e.g., when the user is accessing a Web server internal to the network), the transport service is

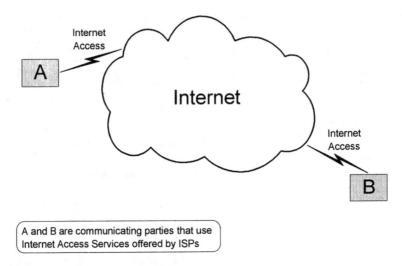

A and B are communicating parties that use
Internet Access Services offered by ISPs

FIGURE 1.3   Use of Internet.

provided by the network, by UDP (User Datagram Protocol), or by TCP (Transmission Control Protocol). IP (Internet Protocol) infrastructure enhancements provide differentiated services at the IP router level. These emerging capabilities of, and extensions to, the current IP features will become key to future applications on the Internet (e.g., RSVP is providing reserved bandwidth to enable Internet telephony and video conferencing). Based on this technology, in particular IP, all types of networks and their services could be standardized. Figure 1.3 shows an Internet example where A and B are communicating parties using the Internet and Internet access services from Internet service providers (ISPs).

This standardization is a threat to proprietary networking architectures, such as Systems Network Architecture, SNA, from IBM. To support both, gateways are being deployed to interconnect both types with each other. It is tempting to consider the Internet as the central switching point of corporate networking. However, performance and security considerations drive corporate network managers to use privately owned Internet-like networking segments, called intranets. Intranet examples are shown in Figure 1.4; A, B, C, and D are communicating parties that use the intranet(s) offered by their own company.

Intranets are company-internal networks that use Internet technology. In particular, Web technology is used for information distribution (e.g., standardized company documentation, internal hiring procedures, etc.) and Web protocols for internal information transfer. The backbone of intranets is IP based on Layer 3. If interconnection is required to other networks (e.g., SNA) or to other companies, then firewalls are deployed to protect the company-owned intranet. Firewalls are actually filters; certain packets without the necessary authorization code cannot pass the firewall.

If partnerships are the targets, networking equipment of partnering companies can be connected to each other. In such a case, the connected intranets are called

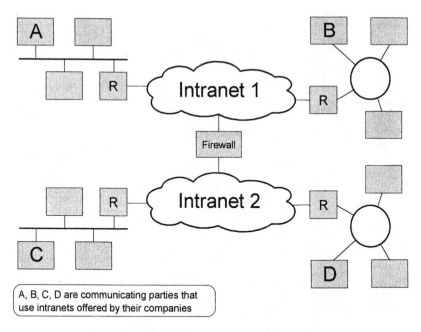

FIGURE 1.4 Use of intranets.

extranets. For such networks, firewall requirements are much lighter. Typical application cases are car manufacturers and their suppliers of parts; airlines in alliance; airlines and travel agencies; telephone companies working with each other to complement local, long distance, and international services; and service providers and customers. Figure 1.5 shows how intranets are connected to each other. Enterprise B is becoming part of Enterprise A by connecting their routers into the intranet.

The Internet can still be utilized as part of intranets and extranets. Virtual private networks (VPNs) are offering this by securing channels that are part of the Internet, to be used by communicating parties in intranets and extranets. There are a couple of technical solutions that are based on either Layer 2 or Layer 3 technologies. Figure 1.6 shows the principles of virtual private networks.

1.3 PERFORMANCE MANAGEMENT CHALLENGES

Feasible network architectures for intranets and extranets are shown in Figures 1.4 and 1.5. The components of the networks are similar to other types of networks. Principal components include:

- Web servers that maintain home pages
- Web browsers that directly support users to view, download, and upload information to/from Web servers
- Backbone offering broader bandwidth for high data volumes
- Access network offering narrower bandwidth for lower data volumes

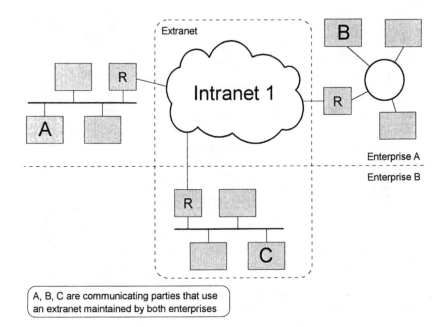

FIGURE 1.5   Use of extranets.

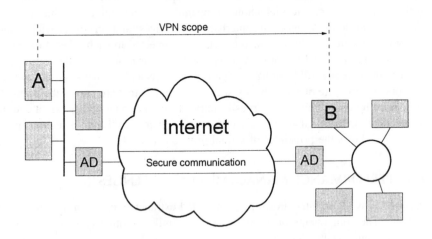

FIGURE 1.6   Use of virtual private networks (VPNs).

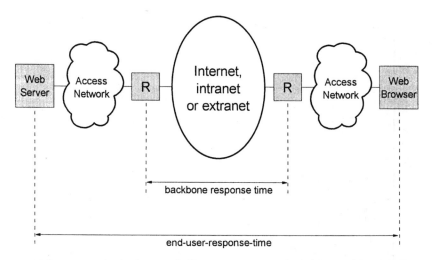

FIGURE 1.7   Principal structure of systems and networking components.

- Networking components including routers, switches, traffic shapers, and firewalls
- Communication protocols such as IP for the backbone and higher layer protocols such as HTTP, SNMP, and FTP to support management applications

Figure 1.7 shows a typical arrangement in a simplified form. From the management perspective, all these components are managed objects. One additional managed object type must be considered; this object type is the application running in Web servers.

In early 1999, approximately 37 million people were accessing the Internet every day. Altogether, approximately 830 million Web pages are accessed every day. Because of these special patterns, performance metrics are extremely important. From the technical viewpoint, everything can be measured. From the practical point of view, however, a few indicators are of prime interest. In particular, two of them are considered in every enterprise:

- Response time
- Resource utilization

For the response time, not only the resource level, but also the user level response time should be measured. There are several types of tools from which to choose: some measure throughput rates, some simulate network traffic and tally the results, some gauge performance by running within the applications themselves, and some rely on a combination of those techniques. Altogether, there are four approaches (Jander, 1998):

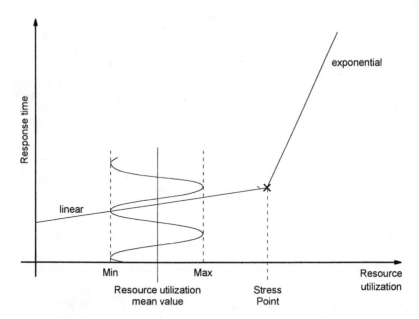

FIGURE 1.8   Response time as a function of resource utilization.

- Monitors or packet analyzers
- Synthetic workload tools
- Application agents
- Application response measurement (ARM) MIBs

End user level response time is helpful for service level agreements. Performance optimization requires more details about the contributors, such as the networks, systems, and applications. When segments of the response time are known, resource optimization by proper capacity planning is possible.

The utilization of resources has a direct impact on the response time. Figure 1.8 shows this mutual dependency. The curve consists of a linear segment until the stress point and an exponential segment beyond the stress point. Webmasters and network managers are expected to determine operational areas of resources. This area is marked in Figure 1.8. Typical stress point level resource utilization metrics include:

- Server CPU for transaction traffic, 80%
- Server CPU for batch-type traffic, 95%
- Router, 75%
- Ethernet, 40%
- Token Ring, 80%
- Communication lines with advanced protocols, 75%
- Communication lines with primitive protocols, 50%

The payload is always an issue with resource utilization. Operating systems put load on the servers; control characters of protocols mean additional bytes to be transferred. Both represent overhead, but they cannot be avoided completely. The same is true with monitors and the transfer of monitored data for further processing. Overhead can be controlled, however, and then productive operations are not impacted. Further details on performance-related metrics in intranets are provided in subsequent chapters.

In summary, tuning and optimizing intranets may be very different than traditional networks. User behavior, application performance, unusual traffic patterns, asynchronous resource demand, and additional protocols create unique challenges to performance management of intranets.

1.4 SECURITY MANAGEMENT CHALLENGES

Due to the fact of opening networks, connecting partners, and using a public domain, such as the Internet, security risks increase considerably. VPNs are one of the possible answers to combine existing infrastructure with acceptable protection. Security expectations may be different in various industries, but generic security management procedures do exist. Security management enables intranet managers to protect sensitive information by:

- Limiting access to Web servers and network devices by users both inside and outside of enterprises
- Notifying the security manager of attempted or actual violations of security

Security management of intranets consists of the following (Leinwand & Fang, 1993):

- Identifying the sensitive information to be protected
- Finding the sensitive access points to sensitive information
- Securing these access points
- Maintaining the secure access points

Identifying sensitive information means the classification of information. Most organizations have well-defined policies regarding what information qualifies as sensitive; often it includes financial, accounting, engineering, and employee information. In addition, however, an environment can have sensitive information unique to it. The main purpose of intranets is to improve the internal documentation and communication within enterprises. Web servers are the focal point of information maintenance. Evidently, not everything is for everyone. Depending on the individual responsibilities, access rights to information sources can be relatively easily structured and implemented. In summary, sensitive information is placed on the home pages with particular content residing on Web servers.

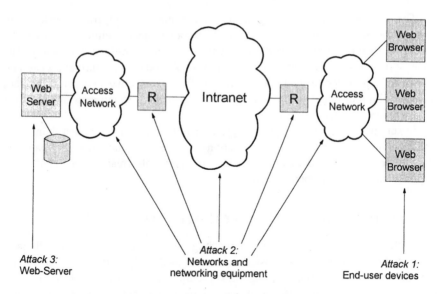

FIGURE 1.9 Access points with security risks.

Once Webmaster and network managers know what information is sensitive and where it is located, they must determine how users can access it. This often time-consuming process usually requires that Webmasters and network managers examine each piece of hardware and software offering a service to users. In this respect, intranets are not different from any other complex networks. Generic sensitive access points include (Figure 1.9):

- End user devices, such as browsers
- Access and backbone networks
- Web servers maintaining sensitive information

1.4.1 SECURITY RISKS IN INTRANETS

Intranets are constantly changing: new Web sites, new servers, new services, and new networking components create new security risks. These risks may be subdivided into high-, medium-, and low-risk categories. High-level risks in intranets include:

- *Brute force attacks*: Many networked machines are shipped with default accounts that allow an administrator to gain immediate access to a machine and to configure it. If the administrator does not change the defaults, an intruder can use them to gain access to the network. When the administrator addresses accounts to a machine, those accounts may be installed with an easy password. A brute force attack against a machine looks for common defaults and known accounts that might be vulnerable. If a default or login account becomes compromised, the services telnetd, ftpd, rsh, and rexec allow access to a machine. An audit tool can perform, through these services, brute force tests for default and vulnerable accounts.

- *Web servers*: The most popular freely available Web server is from NCSA (National Center for Supercomputing Applications). Up to a certain version, it contained a serious vulnerability that allowed anyone from the Internet to run commands through the Web server and gain further access to the machine. Because of the popularity of this Web server, many sites still run this vulnerable version and are open to compromise. An audit tool can locate these Web servers and the administrator can take corrective actions.

- *Anonymous FTP*: This is a service that allows easy transfer of files. The FTP server has many configuration issues. An improper configuration could allow unauthorized access to the rest of the machine. An audit tool can check for these configuration flaws and determine whether the FTP site is vulnerable.

- *Networked file system (NFS)*: An NFS allows many machines to have a virtual hard drive that operates over the network. If improperly configured, an NFS may allow anyone to access this virtual hard drive. An intruder could then copy, modify, and possibly delete critical data from the NFS, and even gain full access to the machine. An audit tool can find misconfigured NFS servers.

- *File sharing:* Windows NT and Windows 95 use a service called file sharing that allows for sharing files between networked computers. Unfortunately, many people do not realize that this may also allow access to their computers by anyone on the Internet connected to intranets. A penetration tool can find misconfigured file-shared machines and allow the administrator to take corrective action.

- *RSH and RLOGIN*: Both rlogin and rsh vulnerabilities give an intruder instant access to the machine. The rlogin vulnerability affects AIX and Linux machines. It allows everyone to rlogin as root without a password. An intruder issuing the command rlogin hostname.com-1-froot sees the login banner and a shell. An audit tool can locate these vulnerable services and enable the administrator to take corrective actions.

- *XWindows*: Many users have xhost+ in their configuration file. This permits access to the XDisplay by anyone, anywhere. An intruder who can access the XDisplay can obtain keystrokes and remotely execute commands as the user running the XDisplay. It is possible to configure the xhost to authorize only certain hosts, but even then any user from those remote hosts can use the XDisplay to compromise data. An audit tool can locate vulnerable XDisplays.

- *Service broadcast*: Computers and host services that announce themselves for general use are prime candidates for attack.

- *Overriding buffers*: A poorly written service may be bombarded with so many requests that it crashes, allowing illegal commands to be sent. Send mail, ftp, finger, and NCSA's first Web server are vulnerable to this problem.

- *Clear text passwords*: If passwords are not encrypted, they can easily be sniffed from the network.

In addition to these high-level risks, there are medium- and low-level risks found on intranets. Medium-risk vulnerabilities are services that gather information from more critical files but do not allow immediate access to the machine. Medium-risk vulnerabilities can allow intruders access to the password file on which they can try to run a password cracker and gain access to an account in the password file.

Low-level risks are network services for information gathering only. For example, the common service called finger allows anyone to find out who has an account on the machine. This is extremely useful for finding e-mail addresses. Intruders use this service to determine whether the administrator is logged into the machine and to identify possible accounts for a brute force attack. An audit tool can gather the same information and try the same types of attacks to attempt to gain access to a machine.

The next step in security management is to apply the necessary security techniques. The sensitive access points dictate how the protection should be deployed using a combination of policies, procedures, and tools. In this respect, the following levels of security techniques must be considered:

- End user devices, such as universal browsers (use of chipcards or chip-keys)
- Access and backbone networks (use of encryption, authentication, and firewalls)
- Web servers (use of server protection, operating systems protection, special tools, and virus protection)

1.4.2 SECURITY PROCEDURES IN INTRANETS

Security procedures include:

- *Entity authentication*: This mechanism allows identity verification by comparing identification information provided by the entity to the content of a known and trusted information repository. This information may take the form of something the user knows, something the user has, or something the user is. For stronger verification, more than one of these characteristics may be required.
- *Access control lists and security labels*: Access control lists are a form of information repository that contain data relative to the rights and permissions of access granted to each authenticated identity known to the system. Security labeling provides a mechanism to enhance or refine the levels of control imposed on a resource or entity. This is done by defining specific controls on the label tag itself.
- *Encipherment/decipherment*: Cryptography is the mechanism used to ensure confidentiality. It is also used quite frequently in complementing other mechanisms to provide total security solutions. Encipherment and decipherment essentially deal with the transformation of data and/or information from an intelligible format to an unintelligible format and back to an intelligible format. This is basically a mathematical process employing keys and algorithms, applying the key values against the data in a predetermined fashion.

- *Modification detection codes and message authentication codes*: Data integrity is supported by the use of some sort of checking code. Three methods of calculating the checking code are in common use: cyclic redundancy check (CRC), modification detection codes (MDCs), and message authentication codes (MACs). A CRC is relatively easy to compute and has typically been used to recognize hardware failures. It is a weak check for detecting attacks. An MDC is computed using cryptography, but no secret key is used. As a result, MDC is a much stronger check than CRC because it is very difficult to find a second message with the same MDC as the legitimate one. However, an MDC has the same delivery requirements as a CRC, in that a CRC or a MDC may be delivered with data by encrypting it using a secret key shared by the sender and recipient. A MAC is cryptographically delivered using a secret key shared by the sender and recipient, so data may be protected during delivery.
- *Digital signature*: In addition to data integrity, nonrepudiation services such as digital signature are becoming more important to many customers. Digital signatures provide proof of data origin (tells recipient who sent the data) and/or proof of delivery (a receipt for the sender).
- *Authentication by special equipment*: Recently, new solutions have been introduced to identify users to their workstations or browsers. The use of chipcards and chipkeys is based on a personalized set of information hard-coded into the chip. However, loss of the card or key may still lead to unauthorized use.
- *Authentication by personal attributes*: In very sensitive areas, personal attributes, such as keystroke dynamics, signature dynamics, voice, color of eyes, hand scans, fingerprints, and the like, may be used as the basis for identification. The cost of these techniques, however, can be very high.
- *Improving data integrity*: This technique deals with solutions based on a checksum computation. The results are used to expand the message that will be sent to the destination address. The techniques are expected to be sophisticated enough not to be broken easily. The original message and the checksum are encrypted together. Also, time stamps and message identification have to be added to help reconstruct the message. Those additional flags may be encrypted as well.
- *Prevention of traffic flow analysis using fillers*: Fillers may be used to fill time gaps between real data transmissions. If both communications can be encrypted together, the intruder cannot recognize any rationale or trend, or any random, periodic, or other pattern, by listening to the traffic. On the other hand, the use of fillers is not unlimited. It may become very expensive, and communication facilities of intranets may be temporarily overloaded, resulting in performance bottlenecks.

The last step in effectively securing access points in intranets is maintenance. The key to maintenance is locating potential or actual security breaches. It requires an ongoing effort of stress testing intranets, assigning tasks to outside professional

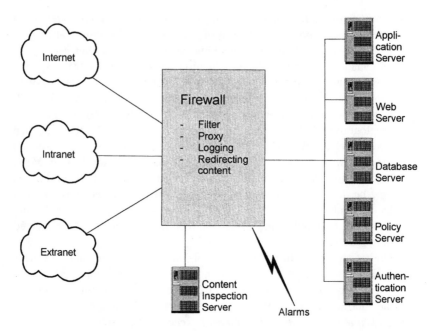

FIGURE 1.10   Firewall architecture.

security companies, reviewing case studies of security violations, and evaluating new security management techniques and tools.

1.4.3   FIREWALLS IN INTRANETS

Firewalls play a significant role in security management of intranets. A firewall (Figure 1.10) is a device that controls the flow of communication between internal and external networks, such as the Internet. A firewall serves several functions. First, it acts as a filter for inbound Internet traffic to the servers of enterprises. It prevents unnecessary network packets from reaching Web and application servers. Second, it provides proxy outbound connections to the Internet, maintaining authentication of the internal Internet users. Third, the firewall logs traffic, providing an audit trail for usage reports and various planning purposes.

Firewalls are not without risks. Many companies assume that once they have installed a firewall, they have reduced all their network security risks. Typically, firewalls are difficult to penetrate, but when they are broken, the internal network is practically open to the intruder.

Furthermore, a firewall does not address internal network compromise. Approximately 70% of all network security breaches occur from within the corporation; that is, by persons already past a firewall. A modem dial-up established by the company or by an engineer for remote access is one easy way past a firewall. In addition, misconfigured firewalls may cause problems. Firewalls are highly susceptible to human error. In a dynamically changing environment, system managers routinely reconfigure firewalls without regard to security implications. Access control lists on a firewall can be numerous and confusing. Intranet managers should be sure that firewalls have been set up correctly and that they are performing well. The following risks have been observed with firewalls:

- *Source porting*: Filter rules are based on source and destination port addresses. A TCP/IP-enabled machine has 65,535 possible virtual ports; some of them are defined for certain services (e.g., e-mail is port 25). When one machine FTPs to another and wants to transfer a file back from the FTP server, typically the server opens source port 20 to connect to the FTP client and transfer data. Therefore, many firewalls allow source port 20 into a network. An intruder can modify telnet to make the connections come from source port 20, thereby penetrating the firewall. A security auditing tool can check to see if source port 20 is allowed to connect to the network.
- *Source routing*: This IP option allows users to define how packets are routed. When source routing is on, many firewall filter rules are bypassed. Many router-based firewalls allow source-routed packets to pass. Many Unix hosts have source routing built into the kernel and do not allow it to be turned off. A security auditing tool is needed to check whether source-routed packets can make a pass through the firewall and connect to various services.
- *SOCKs*: SOCKs is a library of proxy application firewalls designed to allow certain services through and keep intruders out. The fundamental problem with SOCKs is the same as with many security tools: they are often misconfigured. Often the administrator establishes rules to allow certain services through the firewall, but the rules necessary for denying access to intruders are never implemented. Consequently, services seemingly work fine with the firewall, but the firewall's inability to keep intruders out is not recognized until an intruder breaks through. Even then, the cause of the problem may never be recognized. A security auditing tool can attempt to connect to important services through the SOCKs port, to see whether filter rules have been configured properly.
- *TCP sequence prediction*: TCP sequence prediction or "spoofing" tries to trick a host that trusts another host. For example, if Host A and Host B are in a corporate network and Host A is trusted by Host B, then Host A is allowed to log into Host B based on this trust, without a password. Intruders who can make their Host C look like Host A will also be able to log into Host B. A security auditing tool should be used to test for four factors regarding TCP sequence prediction:

    Is the host TCP sequence predictable?
    Does the firewall or router block spoofed packets from the internal network?
    Are the exploited services — rsh and rlogin — running?
    Can the machine be exploited?

- *Direct RPC scan*: The portmapper is a service that allows users to identify the ports on which the remote procedure call (RPC) reside. Many filter-based firewalls may block the portmapper on port 111. The RPC commands themselves remain in place on various machine ports. It usually is difficult to determine where the services are if the portmapper is blocked. However, if an intruder scans directly for the RPC services, he or she could bypass this type of security.
- *Stealth scanning*: In stealth scanning, an intruder does not attempt to establish a connection, but rather uses packets at a low level with the interface. These low-level packets elicit different responses depending on whether or not a port is active. This technique allows TCP port scanning many times faster than a regular connect routine on Unix and does not trigger alarms built into many Satan detectors and tcp_wrappers. Although many firewalls block particular packets that would establish a connection, stealth scanning packets do not attempt to establish a connection; therefore, they can bypass firewall security and identify services running on an internal network.
- *Connectionless protocols*: Firewalls have difficulty tracking packets used for services that do not require established connections, such as the UDP.

For intranets, a network-based intrusion detection system is required to protect the perimeter network from hacker attack. Network-based intrusion detection systems may be deployed as probes or agents running on servers. Probes are the most effective method at providing network-based intrusion detection. A probe minimizes the impact to existing systems because it is a passive listener reporting back to a centralized console without interruption. Intrusion detection will perform, at the network device level, the following functions:

- Inspect data streams as they pass through the network, identify the signatures of unauthorized activity, and activate defense procedures
- Generate an alarm immediately upon detection of the event, notifying the appropriate security personnel
- Trigger an automated response to the several issues to be considered

In addition to intrusion detection, a TCP proxy aggregator may be considered. This will tighten security through the firewall by limiting the exposed ports. It also provides an offload for session/connection management and a more robust technical implementation in terms of port permutations supported.

Tunneling and encryption are used to deploy networks needing to appear point-to-point, but in fact consisting of various routes to an endpoint, providing data integrity and confidentiality. Usually, tunneling protocols, such as Layer 2 tunneling protocol (L2TP), point-to-point tunneling protocol (PPTP), and IPSec (Internet protocol security), and encryption standards such as DES, MD5, Triple DES, and others are used.

Mobile code programs, such as Java and ActiveX, pose an increasing security threat. Content inspection software should:

- Provide full control over Java, ActiveX, and other mobile code activity in the corporation
- Prevent undetected, costly mobile code attacks, such as industrial espionage and data modification
- Enable safe Internet/intranet/extranet surfing while taking full advantage of Java and ActiveX technologies

A content inspection server will accept mobile contents redirected from a firewall to scan for attack signatures. If the scan detects a vulnerability, the contents will be blocked and the client prevented from downloading the mobile code. This denial will alert an appropriate administrator and notify the requesting client. If the scan does not detect any vulnerability, the mobile code is redirected to the firewall for routing to the client.

In summary, security management challenges increase in intranets due to many access points in the network. New techniques and new tools are required in combination.

1.5   ACCOUNTING MANAGEMENT CHALLENGES

As far as the components of intranets are concerned, they are the same as for other types of networks and systems. However, there are fundamental differences in terms of traffic patterns, which may affect the right accounting strategies. Accounting management involves collecting data on resource usage to establish metrics, check thresholds, and finally bill users. Billing is a management decision, but usage measurements are a must in intranets. The principal steps of accounting are:

- Gathering data about the utilization of Web servers, the access, and backbone networks
- Setting usage quotas as part of service level agreements with users
- Billing users for their use of resources

Figure 1.11 shows these typical processes in accounting management.

To gather data on usage, proper instrumentation is necessary. Stand-alone monitors, log file analyzers, and built-in accounting agents are most commonly used. Accounting management requires continuous measurements, but the amount of data collected is usually not critical in terms of overhead. Chapters 5, 6, and 7 describe a number of products that may be used for collecting data on resource usage.

Service level agreements may include an expected level of resource utilization by single users or user groups. Either time duration or byte volumes may be agreed upon. If the agreed data volumes quotas are exceeded, the service and/or the price may change. The agreements and their continuous surveillance help the service provider plan for the right amount of capacity.

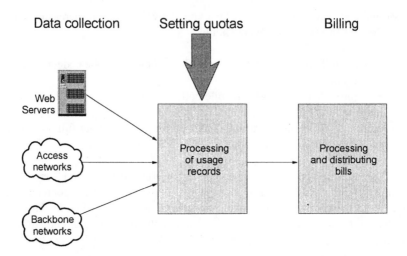

FIGURE 1.11   Accounting management process steps.

Billing with intranet services is a new area, not yet well understood. Users are often billed based on one of the following:

- One-time installation fee and monthly fees
- Fees based on the amount of resources used

The first case is very straightforward. The user is billed for the installation of the intranet access and then a standard fee for each month of use. Using this method, accounting management is not necessary for billing. Although this is the easiest system to implement, it is difficult to justify why users with very different traffic patterns and volumes are billed the same amount.

The second case is more difficult to implement and manage. Again, there are more alternatives, such as:

- Billing is based on the total number of visits
- Billing is based on the total number of packets sent or received
- Billing is based on the total number of bytes sent or received

The accounting and billing cases are more complicated when multiple suppliers are present in intranets. If so, administrators must use a clearinghouse to gather usage data, allocate them to each other, and then generate convergent bills to the users. It is expected that the user receives just one bill for the intranet service.

In summary, the accounting management process can be fundamentally different in intranets in comparison to WANs and LANs of private enterprise networks. In particular, usage-based data collection and convergent billing are the real challenges to accounting management.

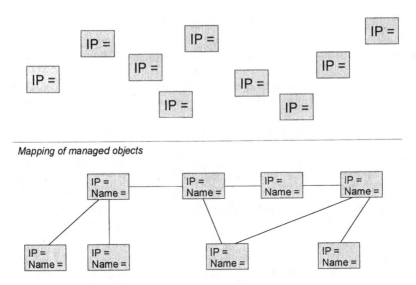

Mapping of managed objects

FIGURE 1.12   Difference between discovery and mapping.

1.6   CONFIGURATION MANAGEMENT CHALLENGES

Configuration management is the process of identifying systems and network components and using that data to maintain the setup of all intranet resources.

Configuration management consists of the following steps:

- Identifying the current intranet environment
- Modifying that environment by moves, adds, and changes
- Maintaining an up-to-date inventory of the components and generating various reports

1.6.1   IDENTIFYING THE CURRENT INTRANET ENVIRONMENT

This process can be done manually by engineers or automatically by intranet management systems. Intranets do not require special treatment. This discovery and mapping step is identical to that performed in other networks and systems. Figure 1.12 shows the difference between the results of simple discovery and mapping. There are a number of sophisticated and powerful graphics tools that can paint and maintain intranet configurations.

SNMP-oriented platforms offer configuration and topology services in two different ways: the discovery function identifies all managed objects with valid IP addresses on the LAN or across LANs, and the mapping function goes one step further by displaying the actual topology of the LAN or across LANs. Both functions can be successfully used for intranets. Managed objects without IP addresses are not discovered. The discovery and mapping processes require time and thus may affect production. Careful selection of the periodicity is required. Many companies deploy intranet visualization tools, instead or in addition to discovery and mapping. Usually, they are very user friendly, but they are independent from the actual network. Without synchronization of the tool's database with the actual network, these visualization tools are useless. However, a combination of the discovery feature of the management platform with a visualization application can be very successful.

1.6.2   MODIFYING THE CONFIGURATION ENVIRONMENT

The intranet environment exhibits an above-average rate of moves, adds, and changes, due to users moving, restructuring buildings and infrastructures, deploying new applications, and the usual equipment changes. To offer service to mobile users, the change rate is not even predictable. Modification would be manual if the data collection method were manual, and automatic if the data collection method were automatic.

Management of moves, adds, and changes requires stable and easily implemented procedures. Intranets become a very important part of the IT infrastructure, requiring high availability and good performance. The moves, adds, and changes window is narrowing with the requirement that they be prepared very carefully. The requester is expected to fill in forms detailing the nature of changes, their impacts on other managed objects, fallback procedures, desired dates, priorities, and human resources requirements. In addition, this process should be carefully monitored. When problems occur, fallback procedures are expected to be triggered. After all the moves, adds, and changes have been successfully completed, all related files and databases must be updated accordingly.

1.6.3   MAINTAINING THE CONFIGURATION

Asset and inventory management is one of the critical success factors of intranet management. Usually, relational databases are used to store and maintain technical and financial data on system and network components. Access is usually via SQL; reporting is supported by standard or additional third-party reporting tools. Figure 1.13 shows a simple example of how a relational structure can be designed and deployed. Object orientation may penetrate this area soon, bringing some benefits in terms of reducing the physical storage requirements.

Asset management is expected to work together with tools implemented in other management areas. In particular, the following links to asset management are obvious in managing intranets:

- Trouble ticketing
- Performance tuning
- Security violations traces
- Accounting details

In summary, managing the configurations of intranets does not introduce additional challenges to configuration management.

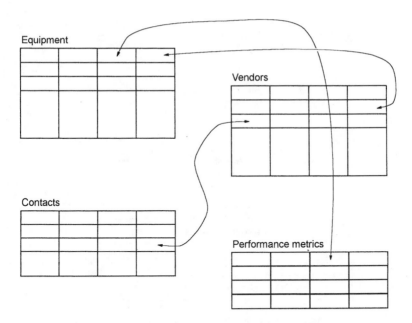

FIGURE 1.13   Connecting relational tables to support configuration management.

1.7   FAULT MANAGEMENT CHALLENGES

Fault management is the process of detecting, locating, isolating, diagnosing, and correcting problems in intranets. Fault management consists of the following steps (Figure 1.14):

- *Detecting and locating problems*: Intranet components generate a number of messages, events, and alarms. Meaningful filtering, combined with user input, helps to detect abnormal operations. Management platforms and their management applications are usually able to determine the location of faults. This phase indicates that something is wrong.
- *Determining the cause of the problem*: Based upon information generated by element managers or correlation results provided by management platforms, the cause of the problem is determined. This phase indicates what is wrong.
- *Diagnosing the root cause of the problem*: In-depth measurements, tests, and further correlating messages, events, and alarms will help to determine the root cause of problems. This phase indicates why the problem happened.
- *Correcting the problem*: Through the use of various hardware and software techniques, managed objects are repaired or replaced, and operations can return to normal. This phase indicates that the problem has been resolved.

In summary, managing faults in intranets does not introduce additional challenges to fault management.

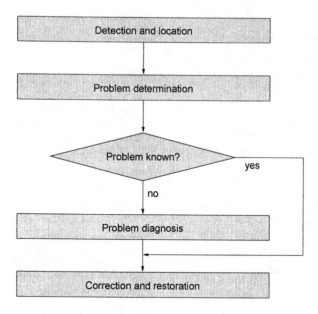

FIGURE 1.14   Fault management process steps.

1.8   THE ROLE OF INTERNET SERVICE PROVIDERS (ISPs)

Intranets represent a dynamically changing network environment. On the way to intranets, enterprises work a lot with ISPs. Even after the deployment of intranets, the working relationships to ISPs remain. Additional firewalls are implemented to separate the networking environment into well-protected enterprise and non-enterprise segments. Managers of intranets must deal with very different service providers, such as (Terplan, 1998a)

- Long distance carriers
- RBOCs (Regional Bell Operating Companies)
- ILECs (Incumbent Local Exchange Carriers)
- CLECs (Competitive Local Exchange Carriers)
- System integrators
- Outsourcers
- ELECs (Enterprise Local Exchange Carriers)
- New entrants specialized in this business

In addition to these service providers, the other suppliers of systems and network components should be managed as well. In this respect, intranets do not differ much from other networking environments. The prerequisites for successful intranets are:

- Service level agreements with each service provider and equipment supplier
- Continuously monitoring performance metrics

- Targeting of strategic partnerships or long-range contracts with providers and suppliers
- Step-by-step, intranet management functions that may be outsourced (e.g., maintenance of equipment, content authoring, hosting home pages, etc.)

1.9 SUMMARY

The typical management processes, such as fault, configuration, performance, security, and accounting, can be successfully reimplemented in intranets. The physical components of intranets are not different from other networks. However, the protocol basis is IP, which offers a standardized way of communication for everybody with a valid IP address. Usage patterns, performance expectations, and security challenges are different in intranets. Intranet management tools are just emerging; the combination of these emerging tools, such as log file analyzers, load balancers, and traffic shapers, with existing management platforms and applications will offer the optimal tool set to Webmasters and Web administrators.

The following chapters will help the reader understand and deploy intranet-specific management processes and tools.

2 Basics of Web Technology

CONTENTS

2.1 INTRODUCTION

This book is not intended to cover Web basics in detail. This chapter reviews a few basic terms that are required to understand how intranets can be managed. It offers a short introduction to Web basics. These basics are absolutely necessary to understand the applicability of management processes and tools to manage intranets.

2.2 WEB TECHNOLOGY

The World Wide Web is an Internet technology that is layered on top of basic TCP/IP services. It is now the most popular Internet application after electronic mail. Like most successful Internet technologies, the underlying central functionality of the Web is rather simple:

- A file naming mechanism — the universal resource locator (URL)
- A typed, stateless retrieval protocol — the hypertext transfer protocol (HTTP)
- A minimal formatting language with hypertext links — the hypertext markup language (HTML)

2.2.1 THE UNIVERSAL RESOURCE LOCATOR (URL)

The URL is part of a larger family of file naming mechanisms called universal resource names (URNs) that are used to designate objects within the WWW. URLs name the physical location of an object; URNs identify without regard to location. Uniform resource citations (URCs) describe properties of an object. At this time, only URLs are in widespread use. URL is the home page address, such as: http://www.snmp.com/.

Like TCP/IP, SNMP, and other popular protocols, URLs were originally considered to be temporary solutions until more powerful mechanisms could be developed. The simplicity and intuitive nature of URLs no doubt contributed to their rapid acceptance.

2.2.2 WEB BROWSERS

Web browsers function as clients, asking Web servers for information by using the HTTP protocol. Each request is handled by its own TCP connection and is independent of each previous request. The retrieval of just one HTML page may require establishing several TCP connections. Consequently, network managers need to be aware of the resource limitations of their Internet or intranet infrastructure when rolling out Web applications, since Web usage is significantly resource consumptive. Examples of popular Web browsers include:

- Mosaic
- MS Internet Explorer
- Netscape
- HotJava
- Webspace

These and other Web browsers are increasingly being used to support internally developed corporate Web applications ranging from company job postings and notices about benefit policy updates to Lotus Notes–based groupware activities.

2.2.3 WEB SERVERS

Web servers store, maintain, and distribute information to clients using the HTTP protocol. They contain the Web pages, which may be individually designed and maintained by the owners of the Web servers. There are numerous hardware and software platforms to support Web servers, including:

- Connect One server
- MS Internet Information Server
- Navisoft/AOL Naviserver
- Netscape Commerce Server
- Open Market Webserver
- Secureware Secure Web Server
- Spry Internet Office Web Server
- Spyglass Server

Server software is needed to handle requests form browsers, to retrieve files, and to run application programs. Web servers must usually handle large numbers of

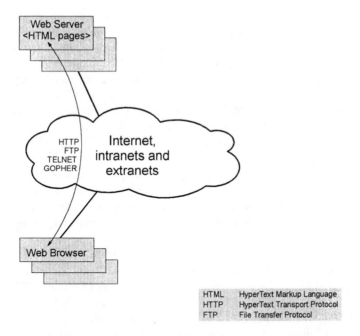

FIGURE 2.1 Web server and Web browser communication.

requests and deal with difficult security issues. The performance of the Web server significantly affects the overall performance of the intranet. Web servers are available for both Unix and Windows NT from Microsoft, Netscape, and others. It is expected that in the next 2 to 5 years stand-alone Web servers will be replaced by servers that are an integral part of the operating system. In addition, these Web servers will handle many tasks that require custom programming today, such as seamless connection to databases, video and audio processing, and document management. Figure 2.1 shows the basic structure of server–browser communication.

2.3 THE HYPERTEXT MARKUP LANGUAGE (HTML)

Web browsers have become widely popular because they all share understanding of a simple media type — HTML formatting language. HTML is easy to understand, and it can be written by hand or generated from other text formats by translators. HTML is actually a simple document type of the standardized generalized markup language (SGML).

HTML is simpler than nroff and other document languages in that it is not programmable. As a result, the descriptive capabilities of HTML are limited to low-level constructs, such as emphasis or indented lists. However, because HTML parsers are rather forgiving of HTML coding violations, many Web pages contain coding mistakes used purposely to achieve particular layout effects on popular browsers.

HTML is optimized for display rather than printing or storage. HTML has no notion of pages, making formatted printing difficult.

HTML has serious limitations. HTML does not provide the flexibility Web publishers need to create home pages. HTML pages are static, and dynamic updates are not really supported. Both attributes, flexibility and dynamics, are absolutely necessary if Web technology is going to be successfully implemented for network and systems management. At one time, most technologies added interactivity to pages by using server-based common gateway interface (CGI) programs, Java applets, browser plug-ins, ActiveX controls, and scripting languages, which had little to do with HTML. Now, however, with dynamic HTML (DHTML), new client-side technologies, combined with scripting languages like JavaScript, may solve many of HTML's problems.

DHTML extends the current set of HTML elements, and a few other elements such as style sheet properties, by allowing them to be accessed and modified by scripting languages. Dynamic features, making pages come alive with movement and interactivity, can be added by exposing tags to scripts written in a language such as JavaScript or VBScript (Powell, 1997).

The tags are accessed through the document object model (DOM). The DOM describes each document as a collection of individual objects such as images, paragraphs, and forms down to individual characters. The DOM of DHTML can be complex, but does not always require much work (Powell, 1997). Developers may use the object model to find an image on a page and replace it with another when a user rolls a cursor over it. Such rollovers, or animated buttons, are common. DHTML also can animate a page by moving objects, build an expanding tree structure to navigate a site, or create a complex application like a database front end.

The common denominator is expected to be the DOM. DOM is the basis for DHTML. It is a platform- and language-neutral interface that allows programs and scripts to dynamically access and update the content, structure, and style of documents.

DOM has been accepted by both of the leading suppliers: Microsoft and Netscape. Their DOM implementation is very similar; differences appear with other features, such as positioning, dynamic fonts, and multimedia controls. With support for CSS and absolute positioning, advanced layout can be made to work under each browser. With DHTML and absolute positioning, it is possible to create sophisticated multimedia applications that can avoid frequent dialogs with the Web server. However, building DHTML-based pages is still programming. Including dynamic elements in a page is a major step away from a static page paradigm and into the idea of Web pages as programs.

The DOM sets out the methods by which Web developers can access elements of HTML and XML documents to manipulate page elements and create dynamic effects, and it serves as the key enabling technology for DHTML. There are three principal areas of XML applications:

1. High-end publishing, which views XML and SGML as highly structured document language
2. Web development of application-specific markup tags, using the extensible nature of XML
3. Distributed Web applications, which use XML as a data exchange format

Despite many benefits of XML over DHTML, users focus on the combination of both. With XML and DHTML, it is relatively easy to share the user interface and information on the Web. The result is that interface development must not be duplicated.

XML can help to eliminate the major limitations of HTML, in areas such as Web searching, fostering inter-industry communication, and enabling a new form of distributed Web-based applications. XML does not solve everything, however, and it does not make HTML obsolete. The origin of both languages is the same (Terplan, 1998a).

Both HTML and XML are subsets of SGML, but XML could define HTML as a DTD (document type definition) of its own. They meet again only with regard to DHTML in which both require the use of DOM.

Although the core syntax of XML is fairly well defined, many other areas need to be addressed. XML provides no presentation services. Another technology must be deployed to present XML data within a Web browser. Eventual use of a style sheet language such as CSS or the extensible style language (XSL) seems likely. Many users implement HTML as the presentation language for XML. To support presentation, Java applets may be downloaded to present even complex data forms. XML mirrors SGML in that it lacks linking capabilities. To eliminate this weakness, the extensible linking language (XLL) is being added to XML. In addition, to support scripting capabilities, there is also a need to connect XML with DOM.

Without presentation, scripting, and linking, XML is limited to being just a data format. However, there are applications, defined as "vertical" for supporting specific industries or "horizontal" for generic use. Microsoft has defined CDF (channel definition format) to push content to selected targets. Open software description (OSD) has been defined in XML to support software installation procedures. Synchronized multimedia integration language (SMIL) is used to define multimedia presentations for Web delivery. In addition, meta languages, such as resource definition framework (RDF), will be defined in the future.

The challenges facing XML are significant. The specifications of associated technologies such as style sheets and linking are not yet complete. XML style sheets that are based on DSSSL will most likely compete against CSS. The linking model of XML is more advanced than HTML, but it is incomplete and too complex. The interaction between XML and DOM needs further clarification.

Industry analysts assume that XML will be used together with HTML. HTML is widely used and getting more powerful with CSS and DOM. XML may add formality and extensibility. Formality allows for guaranteed structure, exchange, and machine readability, which is difficult though not impossible with HTML. Extensibility means the opportunity to create specialized languages for specific applications. Such languages may have significant power within particular intranets or in the area of managing networks and systems.

2.4   TCP/IP AND HTTP

Since HTTP runs on top of TCP, TCP's behavior as a Web transport protocol deserves careful study. TCP provides some inherent flow control via end-to-end acknowledgments and TCP window size adjustments between two application endpoints. This mechanism allows applications to detect packet losses due to varying network conditions and rate-adapts automatically. However, this built-in flow control mechanism was designed originally for long-lived flows over very low bandwidth, long haul pipes. In that scenario, the number of unacknowledged packets travelling through the pipe at any instant in time is small. This limits the number of packets that may have to be resent in the case of packet loss. In situations where TCP packets travel through higher speed links or over longer delay end-to-end paths, the number of unacknowledged packets becomes large, and TCP becomes more susceptible to expensive packet loss recovery. This problem is exacerbated by TCP's congestion avoidance mechanisms, which can result in oscillation between congested and uncongested conditions on busy links. TCP also exhibits behavior, commonly referred to as "greedy source." For long-lived flows, TCP will continue to grow the number of unacknowledged bytes outstanding (or window size) until it occupies all the bandwidth available in the connection. This process, called "slow start," can take several roundtrips to complete. The bandwidth the flow will consume is limited only by the speed of the lowest speed link in the path of the connection with other flows sharing the pipe.

TCP cannot distinguish between congestion caused by a sudden traffic burst at the edge and that caused by true network congestion at the core of the Internet. Instead, it simply assumes the latter in all cases. To deal with burst-related congestion at the edge for short-lived flows, the best strategy is for the networking device at the edge (i.e., between the Web farm and the Internet) to allocate enough buffers to cope with them. For these flows, allocating additional buffers during a burst has the effect of smoothing out packet processing load for the switch without the side effect of overallocating buffers at the expense of other active flows. The explanation for this is as follows: if a flow is short-lived, its net buffer requirements will peak only for a moment. If flows can be classified according to size, bandwidth, and type, it is possible to allocate buffer, switch, and uplink bandwidth resources accordingly.

Most TCP buffers will not be the cure of all TCP problems, however. For long-lived flows, fairness issues and susceptibility to packet loss recovery must be carefully considered. In early ATM switches and in routers, the same conclusion was reached — that is, throwing buffers at it does not solve the problem.

Rate shaping techniques, in which a networking device in the data path intercepts TCP acknowledgments and alters their pacing and advertised window size, show promise, but they suffer scaling limitations when deployed at the termination point for large numbers of flows (i.e., at the uplink for a large Web site). This is due to the requirement to maintain state information in the networking device for each flow. In addition, these techniques provide very little benefit for typical, short-lived Web flows. These flows are often just completing a slow start when the flow terminates. Hence, window management techniques never have an opportunity to kick in. Obviously, these techniques apply only to TCP traffic. Although more research is required, these techniques appear most useful at a slow speed WAN demarcation point such as might be found between an enterprise network and the Internet, where it may be useful to rate limit long-lived flows. In the Web farm, the most useful technique for managing WAN and switch resources is to ensure that sufficient bandwidth and buffers are available to support the lowest speed bottleneck link in the flow path. Admission control procedures can ensure that sufficient local switch, buffer, and bandwidth resources are available for admitted flows.

HTTP is the protocol used for access and retrieval of Web pages. As such, it is widely viewed as the core Web protocol. It is an application-level protocol used almost exclusively with TCP. The client, typically a Web browser, asks the Web server for some information via a "Get request." The information exchanged by HTTP can be any data type; it is not limited to HTML.

HTTP usage has already surpassed that of older Internet access and retrieval mechanisms such as file transfer protocol (FTP), telnet, and gopher. However, these older services often coexist with and are supported by HTTP-based Web browsers.

HTTP is a simple protocol; its clients and servers are said to be stateless because they do not have to remember anything beyond the transfer of a single document. However, HTTP's simplicity results in inefficiency. For a typical HTML page, the client first retrieves the HTML page itself, then discovers there are potentially dozens of images contained within the page, and issues a separate HTTP request for each. Each HTTP request requires a separate TCP connection. HTTP pages are not real time. To retrieve new network status, the user must call up the Web page again. HTTP/Web is only good for monitoring one device at a time. This is why Java is considered necessary for continuous monitoring.

To overcome this multistep process, typical Web browsers may open several TCP connections at once. However, this practice may overload slower speed communication links. HTTP is a textual protocol — all headers are transferred as mostly ASCII text — simplifying the writing of simple browsers.

HTTP needs significant performance improvements. This improvement will come in multiple steps. HTTP 1.0 is the basic link between Web browsers and Web servers. HTTP 1.1 will eliminate some major shortcomings of Version 1.0, and NG (Next Generation) will guarantee that systems communicate with each other without limitations and can exchange "self-describing" data.

HTTP 1.0 misuses TCP/IP by creating new connections constantly, resulting in overloaded communications links and deteriorating performance at high session numbers. Version 1.1 improves network performance and reduces congestion by offering a more controllable caching model, including the ability to specify what is cacheable, how long to keep files in cache, and when to revalidate files. HTTP 1.0 assigns domain names to individual Internet addresses. This means large servers hosting many Web sites require many Internet addresses, resulting in a high administration overhead. HTTP 1.1 allows a single server to support hundreds of Web sites with a single Internet address. HTTP-NG will support applications using new data types (e.g., all multimedia file types in use today), including XML. XML data files include instructions on how the content is organized, so the receiver can accept the data without problems. The HTTP-NG packet will contain information explaining the format of the packet content, allowing Web servers to extend new types of content without requiring a client update.

2.5 NEW DIRECTIONS WITH INTRANET TECHNOLOGY

The latest improvements of HTML have been endorsed by the WWW Consortium. These improvements include:

- Support for style sheets — Users can apply a style sheet written in the simple-to-use CSS language to control color, font, and layout of Web pages.
- Internationalization features — To facilitate a truly international WWW, HTML includes features for rendering text written right to left, as, for example, in Hebrew and Arabic. The LANG attribute can be used with many tags to help the browser display text in a manner appropriate to the language in question. There are also features for specifying the character encoding and language of a document designated by a hypertext link.
- Accessibility features — Some users rely on speech synthesizers or Braille readers when browsing the Web. The latest version of HTML includes features that make the Web more accessible to those who are visually impaired or have other disabilities.
- Tables and forms — There are many new features for creating tables and forms in HTML.
- Scripting and multimedia — There are a number of new features for inserting scripts into HTML and a new OBJECT tag to deal with multimedia.
- Images are tied into the background of a page. Users should be careful when using these extensions to ensure all text is readable and the page remains visually appealing.
- Tables allow the designer to control the display of information on a page to be organized into cells within a grid.

Frames allow the partitioning of the browser's viewable space so that the user can have several independent "window panes" within the viewable space at one time. Each frame may contain its own URL, allowing the user to move among and view several different sites concurrently. Another feature of frames is the ability to display a frozen window. This allows certain information to remain constant on the screen, whereas the user can modify information in the other frames. Information that could typically be placed in a frozen frame would include most items normally placed in the footer and links to the home page or table of contents. Frames are a powerful tool that allow for easier navigation through the site.

The sound feature allows a .wav or .au file to be played when the page is opened. Additional tags allow the sound to be replayed as many times as desired. Consideration for the target platform is essential if using sound because many existing workstations do not have sound cards.

The marquee feature allows a line of text to be scrolled across the page repeatedly. Microsoft provides extensions that allow the designer various attributes for the marquee including background color, alignment, how many times the text should loop, etc.

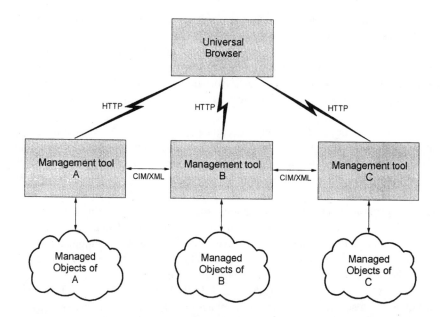

FIGURE 2.2   Loose integration between management tools using CIM and XML.

Microsoft provides support for embedded .wav or .avi files natively within Internet Explorer by extending the standard HTML IMG tag. The most important extension is the dynamic source (DYNSRC) tag, which specifies a video, sound, or VRML world to be displayed. The tag also provides an option for an alternative static image file to be displayed for browsers that do not support video. It is assumed that the browsers support this feature.

Web-based technology can be very successfully implemented to manage systems and networks (Terplan, 1999). A serious shortcoming in current management solutions is the lack of "manager-to-manager" communications, under which management stations from one vendor are able to work with management stations from another vendor. However, a new set of specifications from the Desktop Management Task Force (DMTF) may help to share information between multiple vendors. The model of the new specification is based on the common information model (CIM) and XML. CIM is a conceptual model for describing management information in a format that is independent from a vendor's implementation. XML is adopted as the specification's application-to-application protocol. XML's ability to specify details about elements through DTD provides a way to pass information between different vendors' products.

Figure 2.2 shows how the combination of XML and CIM can be utilized for flexible management and loose integration between management tools. The overall access is via universal browsers; the information to be shared is presented via HTML pages. Element management is the responsibility of the management tool of each particular vendor.

ActiveX is an open platform that combines desktop and Web technologies. This lets developers and Web designers choose from a wide range of language tools, applications, and reusable parts to build rich interactive applications and Web content more quickly and easily. Because there are more than 1000 reusable ActiveX controls already available, Web producers don't have to build all the interactive parts of their Web sites from scratch. ActiveX can be used with a wide variety of programming languages including Visual Basic, Java, Borland Delphi, Visual C++, Borland C++, and others. Developers and Webmasters can make use of their current expertise to create content.

ActiveX can be used for many different types of applications, including small applications delivered over computer networks only when needed. It is compatible with existing investments and enables Web designers to target a wide range of Web users. It can also be used to integrate Web sites with corporate systems and Web browsers with other application software.

Java programmers can take advantage of ActiveX controls directly from Java programs, and ActiveX provides a bridge to Java so that other programming languages from multiple vendors can use Java applets as reusable components. But this way to extend all programming languages, including Java, ActiveX is too new and somewhat unstable at this point in its product cycle.

Besides the popular languages, such as HTML, DHTML, XML, and JavaScripts, two other open source code scripting languages are gaining acceptance among Web site developers: Perl and TCL. Their widespread use is attributable to the fact that they can act as a software integration platform holding together the different elements of the site. Scripting languages are a collection of a few simple commands that are easy to read and understand. Scripting languages are also distinguished by the fact that they are interpreted each time they execute. The more traditional languages, such as C and C++, in contrast, are compiled languages and are compiled at prior execution just once. Interpreted languages may affect the overall performance because interpretation happens during execution. Interpreted languages are good for development, however. Developers can create code and test drive it rapidly, trying out their changes almost as fast as they can make them. C programmers try to make as many changes as they can before submitting their changed code to compilation.

The key target areas of these two scripting languages are (Bobrock, 1998):

Perl

- *Text scanning*: Perl is optimized to scan text files, extract information, and print reports based on that information.
- *CGI scripts*: Perl is the most common language for developing applications that make use of CGI scripts on a Web server.
- *C replacement*: Many of the functions developers built with the C programming language can be done on a Web site, using Perl.

TCL

- *Embedded language:* TCL is a command language that can be embedded in common Web site applications.
- *Tool kit:* TCL comes with the tool kit, which makes it easy to build windowing interfaces in Web applications.
- *Components:* TCL commands can call components from inside applications, integrating Java, C, C++, and other components that cannot otherwise talk to each other.

In the case of Perl, many of the language connections to outside sources of data and application logic already have been written and are available as free, open source code. TCL is a command language; that is, it is meant to issue commands to devices on a network. It can be embedded in the programs of other languages, providing a command structure to control various hardware and software elements.

In addition, the scripting languages have been loosely defined in how they view the data and procedural logic with which they may work. This makes them very flexible and more useful for systems management than a strongly typed language, such as Java. When Java is asked to deal with data from a source it does not recognize, it returns an error message and refuses to accept the data. Perl, on the other hand, tries to hook up to everything it can on a particular system.

Both Perl and TCL have been successful. Their role must be seen as complementary to and not competitive with other Web site development languages.

Digital TV (DTV) is approaching commercial viability. The Advanced Television Engineering Forum (ATVEF) is in charge of creating standards for this technology. The first specification of a standard is based on HTML and JavaScript from Netscape Communications. This specification describes how content providers can create content that can be received anywhere, regardless of where it has been authored, using satellite, over-the-air broadcasting, and cable transmission technologies, and by either analog or digital mode.

2.6 SUMMARY

HTML and its extensions may become the common denominator of future content management applications. In order to go beyond presentation and information distribution, dynamic home page updates are absolutely necessary. Under XLM, with additional language units, the requirements for flexibility and dynamics can be met. Webmasters and administrators do not need to work directly with these underlying technologies. Many tools are available for development, publishing, searching, and site usage analysis. Most of them hide the technical details from users.

3 Intranet Management Challenges: An Overview

CONTENTS

3.1 INTRODUCTION

The emergence of intranets is dramatically altering the way information is accessed within and outside the enterprise. Components of intranets, such as servers, networks, and browsers, are known and are well manageable individually. Their integrated management, however, as intranets, generates several challenges to IT managers. This chapter addresses each challenge and offers guidance to the more detailed explanations of management processes and tools of the forthcoming chapters. Content, server, network, and browser management are all critical success factors. Not giving enough attention to any one of them will cause IT managers to fail in managing their intranets. Figure 3.1 shows the components of intranets.

Web browsers have become the window of choice into corporate documentation and information. There are several important implications of this trend:

- All information can be viewed as Web content, accessible directly through a Web browser, a browser plug-in, or a dynamic piece of code (e.g., Java) that is downloaded automatically to the client. This content can be as varied as a static Web page, a CGI script front-ending an existing database application, or new media such as streaming audio or video.

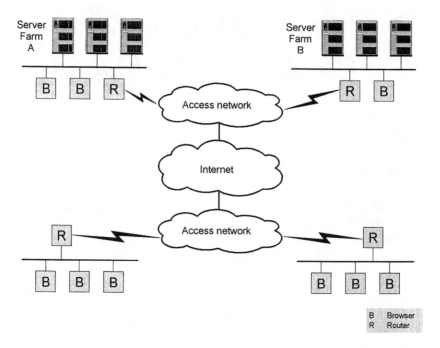

FIGURE 3.1   Components of intranets.

- The information access model has changed from one in which client-specific configuration is required in order to access information to one in which access is always available unless policies are explicitly defined to prevent it.
- Flash crowds, where certain content in the intranet generates significant unexpected traffic, are frequent, making traditional network design techniques based on measuring peak and average loads obsolete.
- Information accessed on or through Web servers comprises the bulk of traffic on the intranet (around 80%). Therefore, effective management of Web resources, bandwidth, and traffic is critical for acceptable quality of service with Web-based computing.

3.2   CONTENT MANAGEMENT

All information can be viewed as content. The way that content is structured and arranged will ultimately determine success or failure. Depending on the content for targeted visitors, page layouts may differ considerably. The content of not only the individual pages but also their links greatly affect visitor satisfaction. Individual visitors expect:

- Readable layout combining text and graphics
- Easy navigation between pages

- Easy return to the home page
- Rapid painting of pages
- Efficient links to interactive services
- Up-to-date status of pages
- Visualization of the site structure
- Site-wide change management of pages
- Easy way of selecting pages to print or download

Goals and interests of companies offering information on home pages include:

- Rationalize information distribution to internal customers
- Fully meet content expectations of external visitors
- Manage intranet resources effectively
- Meet performance expectations of external visitors
- Meet business goals by using intranet technologies
- Provide the opportunity of deploying extranets to link business partners
- Meet high security standards
- Monitor visitors' behavior in order to make rapid changes to increase user satisfaction

Improvements in content management will have a significant positive impact on overall performance. Although Web server performance improvements are part of the performance optimization solution, they must be accompanied by improvements in network and content management technology if they are to have a true impact on WWW scaling and performance. Specifically, developments in the following three areas are critical:

- *Content distribution and replication:* Pushing content closer to the access points where users are located reduces backbone bandwidth requirements and improves response time to the user. Content can be proactively replicated in the network under operator control or dynamically replicated by network elements. Caching servers are an example of network elements that can facilitate the dynamic replication of content. Other devices and models are likely to emerge over time.
- *Content request distribution:* When multiple instances of content exist in a network, the network elements must cooperate to direct a content request to the "best fit" server at any moment in time. This requires an increasing level of "content intelligence" in the network elements themselves.
- *Content driven Web farm resource measurement:* A server or cache in a server farm ultimately services a specific content request. Local server, switching, and uplink bandwidth are precious resources that need to be carefully managed to provide appropriate service levels for Web traffic.

(Chapter 4 is devoted to content authoring and content management.)

3.3 WEB SERVER MANAGEMENT

Web traffic poses a significant number of challenges to existing Internet and intranet infrastructures. Most Web sessions are short-lived. As such, they have fewer TCP packets compared to batch mode operations such as file transfer. In addition, HTTP traffic tends to spike and fall radically. This creates instant demand for hot content, which in turn causes network and server congestions. When Web technology is used to support systems and network management, transport paths are shared between productive traffic and management traffic. Management traffic is even more sensitive against bottlenecks. Web site traffic is highly mobile in that a unique event on a particular Web site could trigger a significantly high hit rate within a very short period. This would be typical in cases with periodic management report distribution and major system and network outages.

Web traffic behavior is significantly different from today's client/server paradigm. It has the following unique characteristics:

- The amount of data sent from a server is significantly larger (5:1) than the amount of data sent from a client to a server. This suggests that optimization of server-to-client traffic has a more significant impact on the intranet and that client request redirection to the best-fit server could have significant performance advantages for Web traffic flows.
- The median transfer size for Web server documents is small (e.g., 5 KB). This implies that Web flows are mostly short-lived flows. They are more likely to create instantaneous congestion due to their bursty nature. Thus, the resource management model must deal appropriately with short-lived flows. Even though HTTP supports persistent connections, due to interoperability issues with existing network caches, it is unclear how widespread deployment will be, or how soon.
- The top 10% of Web server files are accessed 90% of the time and account for 90% of the bytes transferred. This suggests that Web server selection, caching, and content replication schemes that focus on this top 10% will yield the greatest gain.
- A significant percentage (15–40%) of the files and bytes accessed are accessed only once. That is, some small number of large files often consumes a disproportionate amount of total server and network bandwidth. In addition, servers suffer performance degradation when subjected to significant job size variation. This is due primarily to memory fragmentation, which occurs when buffering variable size data in fixed length blocks. Furthermore, subjecting servers to workloads consisting of both hot and one-time requests will result in lower performance due to frequent cache invalidation of the hot objects. Therefore, a server selection strategy that takes into account content, job size, and server cache coherency can significantly improve network and server resource allocation and performance. In addition, requests for large files may be good candidates for redirection to a server that has a shorter roundtrip time to the client.

- Hosts on many networks access Web servers, but 10% of the networks are responsible for more than 75% of this usage. This suggests that resource management strategies that focus on specific client populations may yield positive results in some cases.

Real-time traffic is becoming an increasingly significant proportion of Web traffic. Web site resource management strategies must take into account an increasing demand for support of real-time applications such as voice, distance learning, and streaming media. To deal with both legacy and Web traffic as well as real-time Web traffic, these strategies will need to include admission control as well as bandwidth and buffer allocation components.

The hardware of Web servers is practically the same as with other servers. In most cases the software is divided between Unix and NT; industry analysts expect a clear shift toward NT for price reasons in the future. Besides generic guidelines, Web server sizing should also follow specific criteria determined by analyzing Web traffic patterns. If resource demand is higher than server capacity, multiple servers can be combined into server farms. This solution may satisfy the resource demand criteria, but requires careful attention to allocation and flow control.

3.3.1 CONTENT-SMART QUALITY OF SERVICE (QOS) AND RESOURCE MANAGEMENT

In a typical Web site, the top 10% of Web server files are accessed 90% of the time and are accountable for 90% of the bytes transferred. Therefore, techniques that optimize performance for these files will have the most significant impact on total Web site performance. This requires that the network itself be aware of which content is hot and which servers can provide it. Since content can be hot one instant and cold the next, content-smart switches must learn about hot content by tracking content access history as they process content requests and responses.

Effective management of Web site servers, networks, and bandwidth resources, also requires knowledge of the content size and quality of service requirements. These content attributes can be gleaned through the processing of active flows, proactively probing servers, or administrative definitions. In addition, it is important to track server performance relative to specific pieces of content. All of this information can be maintained in a content database, which provides an analogous function to a routing table in a router or switch. Content-smart switches make a content routing decision based on the information contained in the database, connecting a client to a best-fit server in either a local or remote server farm. This enables the emergence of a business model based on replicating content in distributed data centers, with overflow content delivery capacity and backup in the case of a partial communications failure. Additionally, overflow content capacity intelligence minimizes the need to build out to handle flash crowds for highly requested content.

3.3.2 CONTENT-SMART FLOW ADMISSION CONTROL

Two factors often contribute to congestion in a server farm. One is that servers are not up to the task of handling the amount of incoming traffic. The other is that the link bandwidth from servers to the Internet is overwhelmed by the combination of inbound and outbound traffic; this is complicated by the fact that the amount of outbound traffic from servers is on average about five times of that of the inbound. As a result, a user could make a successful TCP/HTTP connection only to find out that the server could not be allocated the necessary bandwidth to deliver the requested content. To make matters worse, some server implementations come to a grinding halt when presented with an excessive number of TCP/HTTP connections — sometimes requiring a hard reboot. (Chapters 5 and 6 give a number of choices about measuring load and recognizing traffic patterns, and Chapter 7 is fully devoted to Web server management.)

3.4   LOAD DISTRIBUTION AND BALANCING

In order to satisfy the high performance expectations of site visitors, bandwidth in backbone and in access networks must be effectively managed. Usually, servers are consolidated into server farms that are using the infrastructure of a local area network. It is very unlikely that this LAN causes any bottlenecks. Larger enterprises may use multiple server farms deployed at various locations. In order to optimize content allocations, traffic and page references should be monitored and evaluated. At different locations in the network, hardware and software are expected to be installed that intelligently analyze the requests and direct the traffic to the right destination. The "right" destination could be threefold:

1. Server farm destination with the requested content
2. Server farm destination with the lightest load
3. Server farm destination with the closest location to the visitor

There cannot be any compromise on Item 1, but there could be a trade-off between Items 2 and 3, depending on the networking traffic.

The emergence of Web computing and Web traffic over the Internet or intranets has created some unique new problems. It is estimated that over 80% of Internet traffic is related to Web-based HTTP traffic. Even applications such as FTP and RealAudio, which run over TCP and UDP respectively, typically use HTTP to set up the transfer. Since HTTP is an application protocol that runs over TCP, LAN switches and routers, which run Layers 2, 3, and 4, have very limited ability to influence Web traffic behavior. This burden is left to Web servers, which take on the function of TCP/HTTP connection management and, in some cases, the responsibility to distribute HTTP requests to servers within a server farm. This creates inevitable scaling problems as Web sites grow.

The current Internet can be described using a model where local bandwidth is plentiful in the premise LAN located at the edge of the Internet. However, the uplink from the LAN or remote user to the Internet is often severely bandwidth constrained by orders of magnitude. Although congestion can occur anywhere in the Internet path between a client and a server, the most frequent culprits are the WAN connection between the client and the Internet and the WAN connection between the Web farm and the Internet. Actions taken to ensure that this bandwidth is not overcommitted will help improve end-to-end performance.

Instantaneous bandwidth mismatches can occur for a network device that functions as a demarcation point between the public Internet and the Web farm. Examples include:

- The incoming link of the traffic is a faster media type (e.g., fast Ethernet) and the outgoing link is a slower type (e.g., T1 or T3).
- The instantaneous fan-in (i.e., the number of flows being sent at the same time to the same output port) can vary dynamically from one instant to the next.
- A number of traffic sources (e.g., outbound server traffic) may be sharing the bandwidth of a 45-Mbps T3 pipe in a bursty manner over a very high speed switching fabric (e.g., 10 Gbps). This creates a need to regulate flow admission into a slower pipe from multiple higher speed traffic sources.

Information about the use of Web pages, their users, the frequency of access, resource utilization, and traffic volumes can also be collected in the network or at the interfaces of the network. In many cases, the borders between tools and techniques in the server and networking segments are not clear. Tools are different from each other; the differentiators are data collection technologies, performance metrics used, and reports offered.

In the Internet and intranet area, effective bandwidth management is a critical success factor. The role of network planners must be redefined. Real-time and near-real-time bandwidth allocation definitions are needed. Network managers agree that load balancers are needed.

There has been little progress in standardizing load distribution performance metrics. Nevertheless, the following metrics can be successfully used:

- Number of referrals to server farms
- Number of lost requests due to load situations
- Number of requests with an unacceptable response time
- Number of broken connections due to network problems

3.4.1 CONTENT-SMART LINK MANAGEMENT

This technique can ensure that more flows are not admitted than can be handled through the switch or on the uplinks, on average. It is still critical, however, to deal appropriately with traffic bursts and temporary congestion on these links to ensure that Web flows obtain high quality of service. Priority queuing provides a way to prioritize requests based on their type precedence. Fair queuing and weighted queuing methods improve on the priority queuing scheme by addressing the low-priority traffic starvation problem with a scheme that separates traffic into well-identified flows so that each receives a "fair" or "weighted fair" share of transmission bandwidth.

Class-based queuing (CBQ) was developed by the Network Research Group at Lawrence Berkeley Laboratory as an improvement on these existing bandwidth management techniques. It proposes a model in which traffic is categorized in hierarchical classes. Flows inherit their flow characteristics from their parent flow class tree and can have local characteristics of their own. Flows are identified based on the IP address and the inner attributes within the IP header and payload. CBQ provides more granular control of transmission bandwidth and distributes it to member flow classes in accordance with their allocation policies. The model itself is independent of the scheduling techniques that run underneath of it; therefore, implementation details will vary based on the target architecture.

Content-smart link management borrows concepts from CBQ. However, whereas CBQ operates on a packet-by-packet basis based on Layer 3 and 4 classification techniques, content-smart link management classifies flows at admission time based on the content requested, its attributes, and configured policies. These policies support the enterprise and service provider service models described in Sections 3.3.1 and 3.3.2. This facilitates the classification of flows in a two-level hierarchy that includes owners (or customers) and content. Actual scheduling of flows is managed by a hardware-based flow scheduler that supports guaranteed bandwidth flows, prioritized/weighted flows, and best effort flows. Hardware-based scheduling is critical in order to scale the Web farm.

3.4.2   CONTENT-SMART LOAD BALANCING

Simple load balancing techniques such as round robin, weighted round robin, and least connections are inadequate for Web traffic. For example, Web traffic load balancers must support "sticky" connections, which allow a particular server to be selected regardless of server load due to content locality or transaction integrity. Because of the disproportionate ratio of hot content files to total content (1:10), it is highly desirable to support a content replication model that does not require that content be equally and fully mirrored amongst servers in a server farm. This means a load balancing technique must be intelligent enough to recognize if content is available on a particular server before making the selection decision.

Content-smart load balancing takes into account several factors that have a significant impact on the overall performance and cost of a Web server farm:

- *Server cache hit rate*: By directing requests for hot content to a server that has recently serviced that content, ensures that cache hit rate reduces disk access latency for the most frequently accessed content. Because a significant percentage (15–40%) of the files are accessed only once and 90% of the files are accessed only once or not at all, it is important to keep those infrequently accessed files from thrashing a server cache. That is, an infrequently accessed file should be invalidated promptly in server cache to increase the chances that a more frequently accessed file can remain in cache.
- *Burst distribution*: Short-lived, bursty flows can best be handled by distributing them among eligible servers, so long as the servers have been performing below a defined threshold for a period of time.

- *Web flow duration*: Most Web flows are short lived. However, a relatively small number of infrequent, long-lived flows have a far more significant impact on overall bandwidth and server resource consumption. For that reason, long-lived flows should be separated from short-lived flows from a load balancing perspective and short-lived flows of similar QOS requirements should be aggregated to increase TCP flow intensity and reduce per-flow resource allocation overheads.
- *Content-biased server performance measurement*: Current server loading can best be measured by examining the request/response time interval of a server as it handles requests. This measurement is most accurate when connection between the switch and the server is direct. In addition, server performance is not uniform across all content. For example, computer-intensive applications may perform better on one server than another. Other servers may perform better for other types of content. Server performance information needs to be qualified by content.

Load balancers are deployed in different forms. Network managers are confronted with various questions. The questions are:

- Hardware or software based load balancers are better
- Embedded or stand-alone solutions should be preferred
- Use of the combination of both

In the first case, considering high traffic volumes, hardware solutions should be preferred. Software solutions in critical load situations may slow down processes and risk performance. At this time, there are no accurate guidelines for tolerable workload, but a range up to 5% seems reasonable.

Switches, routers, and firewalls are almost everywhere in Internet access networks and in intranets. Embedding traffic control and sharing functions would save extra components but would (as stated earlier) generate additional load and may impair the principal functions. The embedded solution may also include the use of RMON capabilities for real-time load profiling. The stand-alone solution is sensitive against a single point of failure, but would offer overhead-free traffic and load management. The following attributes may play an important role when evaluating alternatives (Rubinson & Terplan, 1998):

Use of load-balancing switches (Figure 3.2)

Benefits:
- Load balancing is performed in a device that is needed anyway in the network
- Centralized management
- Good opportunity to control and guarantee quality of service

Disadvantages:
- Performance may be affected by management functions
- Single point of failure for both switch and management functions

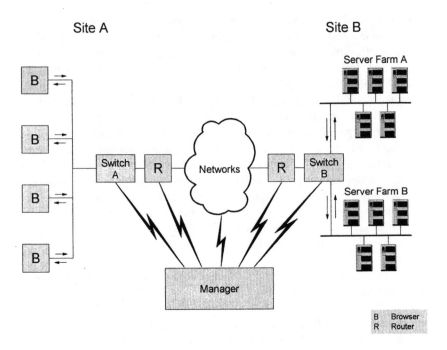

FIGURE 3.2   Use of load-balancing switches.

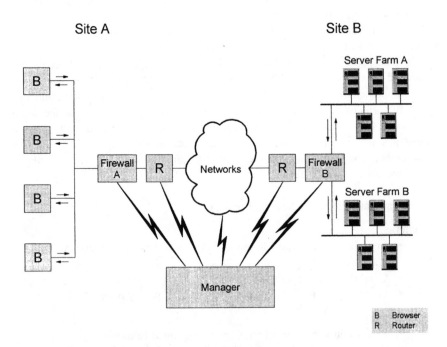

FIGURE 3.3   Use of load-balancing firewalls.

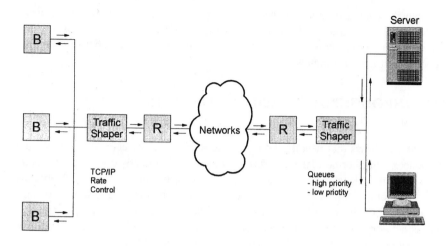

FIGURE 3.4   Use of load-balancing traffic shapers.

Use of load-balancing firewalls (Figure 3.3)

Benefits:

- Load balancing is performed in a device that is needed anyway in most networks
- Centralized management
- Includes special functions and services, such as traffic management and application-based load balancing

Disadvantages:

- Switches are still needed
- Single point of failure for both firewall and management functions
- Performance depends on hardware and operating system configuration

Use of load-balancing traffic shapers (Figure 3.4)

Benefits:

- Load balancing is performed by a device most likely present in the networks anyway
- Centralized management
- Offers traffic shaping and balancing for Internet or intranet access in addition to server access

Disadvantages:

- In most cases, switches and firewalls are needed in addition to these devices
- Single point of failure for both traffic shaping and load balancing
- Little experience yet with performance and scalability

(Chapter 8 focuses on load balancing and on the optimal distribution of load. For performance assessments, Chapter 9 covers useful look-through measurements. These measurements concentrate on the end-to-end response time between communicating partners in intranets.)

3.5 THROUGHPUT CAPABILITIES OF ACCESS NETWORKS

Experience shows that usually the access networks of intranets cause performance bottlenecks. Furthermore, these parts of intranets are frequently not under the control of users, but of ISPs. This critical bandwidth must be managed carefully because user satisfaction and low operational expenses are competing against each other. If the user is in full control, then selection of the technology and proper sizing of access networks are the challenges. Most likely, one of the following technologies is expected to be chosen by ISPs or by users (Terplan, 1998a):

- Dedicated T circuits (Figure 3.5) — The T1/E1 carrier systems are high-capacity networks designed for the digital transmission of voice, data, and video. The original implementations digitalized voice signals in order to take advantage of the benefits of digital technology. The term T1 was devised by the telephone companies to describe a specific type of carrier equipment. Today it is used to define a general carrier system, a data rate, and various multiplexing and framing conventions. A more concise term is DS1, which describes a multiplexed digital signal that is carried by the T carrier. Typical rates are:

DS1	T1	1.544	Mbit/s
DS2	T2	6.312	Mbit/s
DS3	T3	44.736	Mbit/s
DS4	T4	274.176	Mbit/s

Europe and Japan use different throughput rates, but that does not change the basic characteristics of this technology. Upstream and downstream may or even should be sized for various bandwidths.

- ISDN (Figure 3.6) — The initial purpose of ISDN was to provide a digital interface between a user and a network node for the transport of digitized voice and data images. It is now designed to support a wide range of services. Basically, all communication forms may be supported by ISDN. It has been implemented as an evolutionary technology of a telephone-based integrated digital network. Many digital techniques seen with T1 and E1 are used in ISDN. It includes signaling rates, transmission codes, and physical plugs. This technology should use a different multiple of the fixed bandwidth of 64 Kbps. It could even be the difference between basic and primary rates that satisfies the upstream and downstream needs.

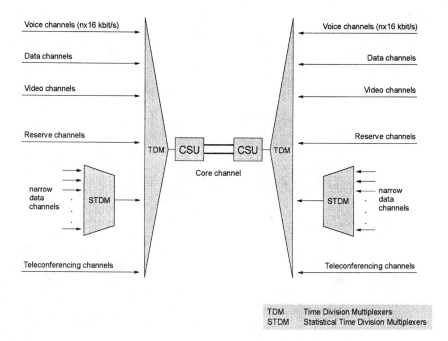

FIGURE 3.5   Configuration of T/E implementations.

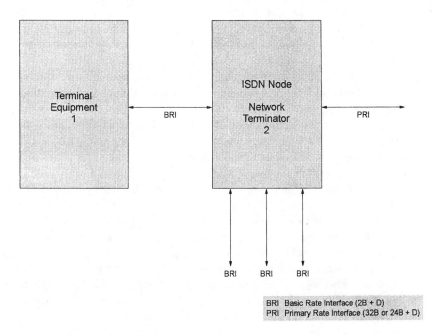

FIGURE 3.6   ISDN basic and primary rates.

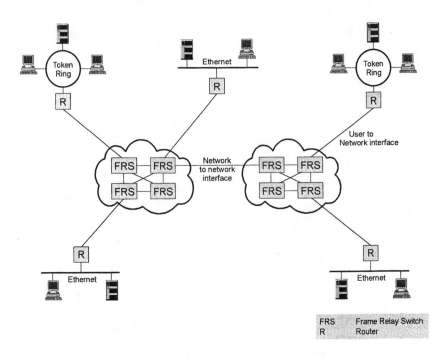

FIGURE 3.7   Structure of typical frame relay networks.

- Frame relay (Figure 3.7) — The purpose of a frame relay network is to provide an end user with a high-speed virtual private network (VPN) capable of supporting applications with large bit-rate transmission requirements. It gives a user T1/E1 access rates at a lesser cost than can be obtained by leasing comparable T1/E1 lines. It is actually a virtually meshed network. The design of frame relay networks is based on the fact that data transmission systems today are experiencing far fewer errors and problems than they did decades ago. During that period, protocols were developed and implemented to cope with error-prone transmission circuits. However, with the increased use of optical fibers, protocols that expand resources dealing with errors become less important. Frame relay takes advantage of this improved reliability by eliminating many of the now unnecessary error checking and correction, editing, and retransmission features that have been part of many data networks for almost two decades. The primary use of this technology is data. If performance metrics can be met, its use can be financially beneficial.
- ATM (asynchronous transfer mode) (Figure 3.8) — The purpose of ATM is to provide a high-speed, low-delay, multiplexing and switching network to support any type of user traffic, such as voice, data, or video applications. ATM segments and multiplexes user traffic into small, fixed-length units called cells. A cell is 53 octets, with 5 octets reserved for the cell header. Each cell is identified with virtual circuit identifiers that are

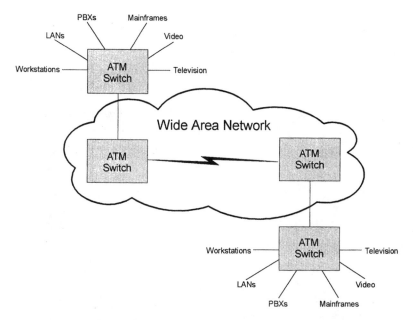

FIGURE 3.8   Typical use of ATM switches in WANs and LANs.

contained in the cell header. An ATM network uses these identifiers to relay the traffic through high-speed switches from the sending customer premises equipment (CPE) to the receiving CPE. ATM provides limited error detection operations. It provides no retransmission services, and few operations are performed on the small header. The intention of this approach — small cells and with minimal services performed — is to implement a network that is fast enough to support multi-megabit transfer rates. The bandwidth reservation can be different for downstream and upstream. Its applicability depends on the specific environment of the user or of the ISP.

- Cable (Figure 3.9) — Cable service providers can now enter the competition for voice and data services. Depending on the country, there are millions of households and businesses with cable television connections. In the majority of cases, cable television is a distribution channel supporting one-way communication only. However, with cable modems, channels could be provided for two-way communication, allowing consumers to send back data or use the cable for phone conversations. There are practically no bandwidth limitations. The nature of this technology is distribution — in other words, downstream. In this respect, it fits well into the Internet philosophy.

- xDSL (digital subscriber line) (Figure 3.10) — The enabling technology is digital subscriber line (xDSL), a scheme that allows mixing data, voice, and video over phone lines. There are, however, different types of DSL to choose from, each suited for different applications. All DSL

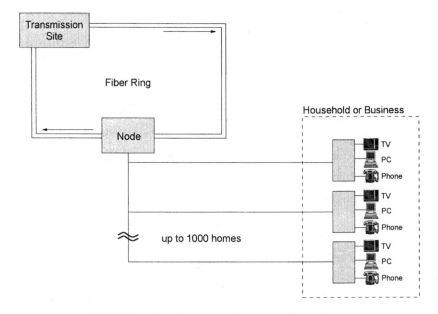

FIGURE 3.9   Cable access for multimedia services.

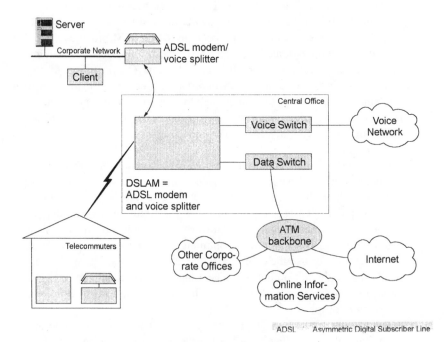

FIGURE 3.10   ADSL in use for multimedia transmissions.

TABLE 3.1
Comparison of Technologies

Criteria	T Circuits	ISDN	Frame Relays	ATM	Cable	xDSL
Suitability	medium	good	good	excellent	excellent	good
Maturity	high	high	high	medium	low	low
Scalability	good	medium	medium	excellent	medium	good
Distance limitations	none	none	none	none	some	high
Costs	high	low	medium	high	low	low

technologies run on existing copper phone lines and use special and sophisticated modulation to increase transmission rates (Aber, 1997). Asymmetric digital subscriber line (ADSL) is the most publicized of the DSL schemes and is commonly used as an ideal transport for linking branch offices and telecommuters in need of high-speed intranet and Internet access. The word asymmetric refers to the fact that it allows more bandwidth downstream (to the consumer) than upstream (from the consumer). Downstream ADSL supports speeds of 1.5 to 8 Mbit/s, depending on the line quality, distance, and wire gauge. Upstream rates range between 16 and 640 Kbit/s, again depending on line quality, distance, and wire gauge. For up to 18,000 ft, ADSL can move data at T1 using standard 24-gauge wire. At distances of 12,000 ft or less, the maximum speed is 8 Mbit/s. ADSL delivers a few other principal benefits. First, ADSL equipment being installed at carriers' central offices offloads overburdened voice switches by moving data traffic off the public switched telephone network and onto data networks — a critical problem resulting from Internet use. Second, the power for ADSL is sent by the carrier over the copper wire with the result that the line works even when the local power fails. This is an advantage over ISDN, which requires a local power supply and thus a separate phone line for comparable service guarantees. A third benefit is again over ISDN: ADSL furnishes three information channels — two for data and one for voice. Thus, data performance is not affected by voice calls. Rollout plans are very aggressive with this service. Its widespread availability is expected for the end of this decade. If distance limitations are eliminated, it can be used very well for intranet access networks. However, this technology is not yet mature and is expected to be used for residential customers only.

Table 3.1 compares the features of these technologies using various criteria. Altogether, there is no clear winner at this time.

3.6 BROWSER PERFORMANCE

In a fully managed environment, browsers should also be included. It is very unlikely that they are to blame for performance bottlenecks. Nevertheless, certain attributes should be carefully evaluated. All experiences with desktop management can be used here.

The performance of a browser has more impact on both the perceived and the actual system performance than any other component. The operating systems are frequently the same or similar from the same supplier as with Web servers. In these cases, coordination between the two is much easier. When the Web browser is running on an old PC with limited RAM, performance may be affected. The machine cannot accept and display data as fast as the Web server and the network are supplying it. At times, it is cheaper and more practical to upgrade the browser by adding more RAM or a coprocessor. The protocol software can also affect the performance. A full seven-layer stack requires considerable resources to run. Even with more user-friendly protocols, performance problems may occur depending on how they select packet sizes, transfer buffers, and translate addresses. The browser executes the network's protocols through its driver software; a faster browser will add to the performance. One factor to consider is whether the browser should contain a disk drive of its own. Obviously, a diskless browser will ease the budget and improve security somewhat. However, diskless browsers have their own set of costs. For one, these browsers depend on shared resources. If the work being performed at the browser does not involve sharing resources, a browser with its own disk may be more appropriate. Moreover, diskless browsers add to the network traffic.

In a Java environment, the browser should be able to accommodate all the applets transferred from the Web server for execution locally.

3.7 SUMMARY

Intranets use the same networking components as any other types of standardized client/server or legacy types of communication networks. In intranets, the lean clients are the universal browsers, offering unified access to information maintained on Web servers. Web servers offer the opportunity to access other servers, such as database and application servers, and convert their content into HTML or XML. All networking components in backbone and access networks are responsible for relaying traffic between Web servers and Web browsers. In terms of sizing and optimizing resources, there are fundamental differences due to the fact that usage patterns are new and altogether not predictable. Resource sizing and optimization are driven by content. The prerequisite for successful operations is the careful analysis of site usage and trend changes in user behavior.

4 Content Authoring and Management

CONTENTS

4.1   INTRODUCTION

Authoring tools present a stand-alone environment in which to build pages. Although this requires learning a new program specifically for HTML creation, these tools allow users to make the most of HTML, using features that traditional word processors do not support.

Currently, there are two distinct kinds of tools Web authors can use to bring their words to the Web. Tag-based tools automate HTML syntax, allowing users to see and tweak tags without having to enter their syntax manually. In contrast, WYSIWYG tools hide HTML from the user, generating it in the background instead. If these tools do not support a specific feature of HTML, that feature must be added manually after the document's underlying code is visible, usually in a text editor. Some products use dialog boxes or palettes to accept information before displaying it as HTML code in the body of the document. Because these tools generate HTML for users, they minimize the learning curve for new Web authors and can produce syntactically perfect HTML. Many of the publicly available tools have both standard and professional features, the latter being available only in the registered or commercial version.

4.2   DESIGN OF HOME PAGES — CONTENT AUTHORING AND DEPLOYMENT

Most users are challenged by the task of information creation, management, and dissemination. These activities are time consuming and difficult to control. The Internet and intranets alone cannot solve information management problems unless specific intranet solutions are implemented that directly address the need for document management. The new discipline, called content authoring and deploying, includes the following tasks (Rubinson & Terplan, 1998):

- Creating content
- Reviewing content
- Approving content
- Changing content
- Deploying content

Figure 4.1 shows the process of creating, reviewing, changing, enhancing, approving, and deploying home pages.

The prerequisites to successfully execute these tasks are:

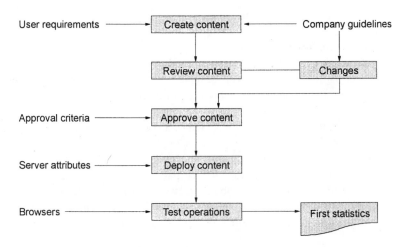

FIGURE 4.1 Process of content authoring and management.

- Users must be able to easily add and update content on a periodic basis.
- Users must be able to protect their page contents from changes by other users.
- A content approval process should be defined and in place. This process should encompass ways to manage and control document revisions, especially changes to shared documents.

As policies and procedures relating to content management are formulated, it is important to designate responsibilities to specific individuals to ensure that they are properly implemented and followed. An internal style guide should be developed that provides page layout, design elements, and HTML code guidelines. Usually, case tools are also involved. The style guide will help users maintain a consistent look and feel throughout the Web pages. Sometimes television-like techniques are helpful in this respect. The style guide should contain information on where to obtain standard icons, buttons, video, and graphics, as well as guidelines on page dimensions and how to link the pages to each other. As part of the style guide, it is helpful to create Web page templates. These templates consist of HTML files and are used to provide a starting point for anyone interested in developing Web pages or content for the intranet. It is fairly simple to create a working Web page and to publish for mass viewing. The real challenges are as follows:

- To maintain the page
- To size the Web server
- To configure the access network

4.2.1 SITE DESIGN CONSIDERATIONS

Content authoring includes a number of tasks. The most important tasks for site design are:

- How to structure a Web site
- How to layout a Web page
- Ideas for improving usability
- Technical hints to increase display speed
- Collection of examples of well-designed sites that can be used as models
- Consideration of new Web technologies for site design

One of the principal factors in the design of a good Web page is knowledge and understanding of the motivations and goals of the target user(s) as well as the technical platforms on which they operate. Given the varying levels of user knowledge and the infinite number of ways a Web page can be constructed, this understanding is essential to creating a usable, effective Web site. Therefore, a user- and task-centered analysis should be completed prior to the design phase to gain knowledge about the target users and their goals. Important questions include:

- To whom will the page be available?
- What are the business drivers for the site (e.g., to provide information, to collect data, to market products)?
- Who are the users (e.g., professional "knowledge workers" or casual intranet users)?
- How will a typical user access the page (e.g., fast connection or dial-up)?
- What browser will they use?
- What are the most frequent tasks that users perform?

Answers to these questions will provide the necessary background information for the navigational structure of the site. During site design, designers should keep in mind that if users cannot quickly find what they are looking for and are not engaged by the layout and information contained within the site, they are likely to move on.

4.2.1.1   Site Registration

The purpose of site registration is to establish content ownership and facilitate navigation. Through the site registration process, sites are added to the intranet directory and become accessible via the intranet-wide search facility. A Web site is defined as a collection of related Web pages. Typically, a Web site is an administrative unit.

4.2.1.2   Site Navigation

There are two points to consider when constructing the navigation layout for a Web site — namely, the structure of the information and how access to that information will be provided. First, the layout of the site is usually the most difficult part of the site design process, particularly if a lot of information will be accessible from the site. Adequate time must be put into designing the structure of the information as to allow easy access for all users. Second, navigation tools must be clear and easy to use as well as functional within all types of browsers that will be used by the target audience. Navigational design must consider the same factors as many other graphical user interfaces (GUIs). Because movement within a Web site is typically nonlinear, navigational menus should be planned to allow users quick access to any part of the site.

4.2.1.3 Content Organization by Menus

Users' ability to move through a Web site and find the information or functions they are searching for plays an important role in determining how successful a site is. Menus and submenus are powerful tools in the design of a Web site. In the same way that menus are used in traditional Windows-based design, HTML menus can be used to subdivide and group relevant content to guide users to their topic of interest gradually. The use of more than four levels of menus forces users to work too hard to find the information they are looking for. Using too few levels may be equally difficult to navigate, in particular when the information volumes grow. Three to four levels should generally provide appropriate depth and guidance for the user. However, because of the varying content of sites, this is a flexible guideline. The menu structure for the site should be continually evaluated and improved as the site grows.

4.2.1.4 Interaction Models

There are many ways to organize information contained within a Web site. The term "interaction model" refers to the structure that is implemented to allow the user access to the various pages within a site. The type of model best suited to a particular page will depend on the content and complexity of the information that the page presents. There are a number of interaction models in use. These models may be used independently or in combination throughout a site. These models are:

- Table of contents — This approach is taken from printed books. Users can easily find the headings they are looking for and then hyperlink directly to that page. This type of access is useful for sites that provide textual or encyclopedic information.
- Image maps — These are graphics that use an embedded linkage map that relates hot spots on the graphic to URLs within the Web site. The user is thus able to point and click on the graphic to move to different locations on the site.
- Graphic menus — These provide the same visual approach to site navigation as image maps without incurring the disadvantages of employing one single large graphic mapped with links. They employ smaller, simpler graphics, strategically placed to provide visual impact.
- Search — Web site searches provide a useful means of allowing a user to access information contained on a particular Web site. Some form of search facility is usually a requirement for larger sites.

- Indexes — These provide functionality similar to book indexes. They allow a user to rapidly locate information pertaining to a specific key word or topic. An index may be used in combination with a search.

4.2.2 PAGE DESIGN CONSIDERATIONS

The actual layout of a Web page is highly dependent on the type of information being presented. This section provides some fundamentals of good page design.

4.2.2.1 Header

The header provides a user with access to commonly used functions within the company-wide intranet and clearly differentiates intranet content from Internet content. The standard header provides navigation links via the following graphics:

- Company logo — links to the company home page
- Directory — links to the company's intranet directory Web site
- Services — links to the company's intranet service page
- Search — links to the company's search Web site
- Help — links to the company's intranet Help Web site

Pre-imaged mapped versions of the company's header should be available on the intranet development and support site.

4.2.2.2 Footer

The footer gives the user important information about the page and provides consistency within the company's intranet. The standard footer usually contains the following:

- A horizontal rule as a separator
- Copyright statement
- Statement regarding content ownership, with an optional e-mail link to the person responsible for maintaining the page (the individual's name should not be used here)
- Date of the latest revision

4.2.2.3 Page Size

Page size must be designed with the actual usable space of the browser window in mind. Typically, this would be the smallest amount of usable space for the standard browser configuration in a 640 x 480 video monitor resolution. Designers should limit horizontal scrolling as much as possible. Keeping the width of the Web site to less than 600 pixels (using tables) makes it much easier for users to navigate information. In some cases, horizontal scrolling is normal and acceptable.

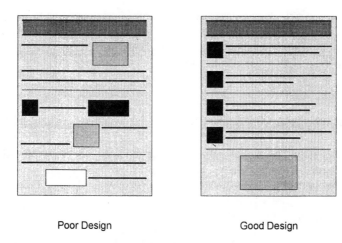

Poor Design   Good Design

FIGURE 4.2   Layout examples.

The acceptable size for an intranet page is 100,000 bytes or less. This includes imbedded images. This size will keep performance within acceptable limits for LAN, WAN, and dial-up users with 28.8-Kbps modems.

4.2.2.4   Home Page

The layout and design of the home page of any Web site are extremely important. Besides being the first thing a user sees upon entering a site, it defines the organizational structure and tone for the entire site. Some essential elements for every home page include:

- Visually appealing design
- Overview of site content
- Links to other parts of the site
- Company/organization identifying information

4.2.2.5   Page Layout

HTML does not provide graphic designers the flexibility they are accustomed to with existing page layout and editing programs (e.g., MS Word and Adobe Page-Maker). However, this does not mean that complex and functional applications cannot be created using HTML. Rather, one must realize that, when used inconsistently, the graphic and typographic controls of HTML can result in inconsistent designs.

To avoid a haphazard look, designers should take care in how graphics are placed and organized. A consistent style will also allow for a consistent conversion from non-HTML documents. It is better to use simple icons and images, instead of complex ones. Navigation icons should be kept in a consistent place. In Figure 4.2, the document on the left is inconsistently organized and difficult to read. The document on the right shows structure, an organized grid approach to design that is implemented consistently across the pages. This will aid users in finding the information they desire, and build their confidence that they are navigating through a well-organized collection of information.

4.2.2.6 Text Style

Text needs to be short, concise, and well organized. When browsing, visitors tend to scan rather than read. They appreciate sections that are ordered logically. Similar ideas or facts should be presented in a consistent way, with the same components presented in the same way in the same order. Consistency is a very important consideration in Web design.

4.2.2.7 Graphics

Graphic images should be used where appropriate to help the user navigate and find information more quickly. Graphics also provide a "look" to the site that will help the user identify where they are. Graphics should not be overused for internal publishing applications. Whereas external marketing Web sites often are graphically intense to catch attention, use of graphics in internal Web sites should be based on ease of navigation and usage. The type, sizing, and location of graphics throughout a site should be presented in a consistent manner — items of similar importance should have the same size and type of graphic. If a larger-than-normal graphic is used, the user is likely to assume that there is some additional significance. Often, the visibility and intended use of the site will dictate the level of graphics required for the site.

Graphic images should be designed for a 256-color environment. A common mistake that professional graphic designers make is designing with higher resolutions and greater color depth than the deployment environment. A color scheme designed in 16-bit color may look bad in 256-color or even worse in 16-color environments. Design should follow the requirements of the target environment.

Most images are between 10 and 30 K. The exception would be image maps on navigational pages or photographic images, which should be around 50 K. One of the drawbacks with using images on a network is the time it takes to download very large files. Images must be kept as small as possible and must fit within the size of the browser's viewable space.

For images, file formats are the best to use. GIF and JPEG are both compressed formats. GIF format is better for smaller graphic or line art images.

4.2.2.8 Local Navigation Elements

Each Web site should include a site map, showing a detailed layout of the site with links to all possible sections and documents. Each page within a Web site should include a link to the site map page. Users may link to a Web site or Web page from a number of different places (navigation page, search results page, hyperlinks, etc.). The site map page gives users a quick and easy way to locate the information they need. On long pages, users may want to quickly go to the top of the page to view the table of contents or other introductory information. The Top of the Page icon helps users more quickly navigate to the top of the current page.

4.2.2.9 Links

Although many Web sites incorporate graphics to support navigation, text links still play an important role in the usability of a site. Working with text in HTML is easy. However, because it is easy to create links and change font types, there are several mistakes commonly made. The following guidelines can help ensure a site's readability and usability:

- Design for scanners, not for readers
- Explain the page's benefit in the upper half of the page
- Use bold typeface to draw attention to a particular section
- Avoid typing in all caps — it is more difficult to read
- Make sure links are underlined in addition to being colored, to assist users who may be color blind or are using black and white monitors
- Avoid blinking text — it is difficult to read and annoying to users

A typical Web page provides both informational text and links to more specific information. Most people look for visual clues to whether a page is useful or interesting enough to be worth reading. If they don't find what they want quickly, they will move to another site. One of the difficulties in using text for navigational purposes is the wording of the links. Proper wording of the text allows users to jump to a new topic or continue reading without losing their place.

All links to default pages should be set with a trailing "/". This eliminates the problem of DNS names turning into IP addresses. By default, the Web browser converts any hyperlink that does not include a Web page (such as a link to a home page) to the default page for the server. However, depending on the browser, this may convert the DNS name into the physical IP address of the hosting server. If the DNS name is converted to an IP address and the user bookmarks the page, the URL will be stored with the IP address. If the IP address of the site changes, the bookmark will no longer work. To eliminate this problem, simply include a trailing "/" on any link that does not include a page file name.

Abstracts and summaries are very helpful for large pages or large graphics. Whenever possible, users should have the opportunity to link to further information if desired. Very large files or files that are not in a usable browser format (e.g., ZIP files, BMP files, etc.) should have a link that allows users to download the file to their local PC.

4.2.2.10 Other Graphic Elements

Separators are graphic or possibly textual elements used to break up or visually divide the contents of a single Web page. Separators can be as simple as a horizontal line, a shadowed line graphic or an actual image file. Their use helps to visually vary subject matter on the page. Although separators can be effective, it is important to remember that they should not distract the user from page content; rather, their purpose is to divide the information into logical groupings.

HTML provides tags for standard information-gathering controls such as radio buttons, dropdown menus, and exit boxes. In general, the guidelines created for traditional GUI-based development apply to Web page design:

- In most languages, the written word is read from left to right; therefore, text should be left-aligned.
- Exit boxes should be similarly sized and also left-aligned.
- Tabs should move the user downward through the page.
- Controls should be evenly spaced and aligned when possible.
- A default button should be provided.
- Mixed-case text (rather than all caps) should be used.

Bullets are used in HTML in the same manner as in traditional word processing, to define a list of items. Although regular bullets are fine, many graphic bullets can add a touch of color to an ordinary Web page.

4.2.2.11   Background and Text Colors

Appealing backgrounds and text colors can add an artistic look to Web sites, but the way colors are used also affects the usability of the site. Designers must be aware that colors have different meanings for different people and that some users may be unable to distinguish some colors.

Some user interface guidelines that are applicable to Web sites include:

- Color is second only to movement in attracting attention.
- Three colors are sufficient for a color scheme.
- Specific colors should be used carefully.
- Shades of red attract attention, but the retina responds to yellow the fastest.
- Blue is difficult to focus on, making unsaturated blue a good choice for backgrounds.
- Gaudy, unpleasant colors and combinations of red/green, blue/yellow, green/blue, and red/blue should be avoided.
- If a background is used, it should be either a light-colored pattern or a solid color.

4.2.2.12   Printing

When the nature of a site is documentation, users must be able to print individual Web pages or an entire site's content. This can easily be accomplished by adding a link to a printable form of the entire document. Documents may also be provided in multiple formats to accommodate the maximum number of users.

4.3   ISSUES WITH CONTENT AUTHORING

The recommendations for Web page design can be summarized as follows:

- Use standard links.
- Use one or more interaction models, such as a table of contents, image maps, graphic menus, a search engine, and indexing.
- Incorporate navigation elements.
- Segment long documents into small ones.
- If a site includes a significant amount of pages or data, provide a local search page to search the content only.
- Design pages for rapid and slow search.
- Use text pages with a narrow bandwidth.
- Test HTML pages and links before practical use.
- Test content on different browsers.
- Use Webmaster's recommended templates to create new pages.
- Use abstracts or summaries for larger text pages or large images and give users the option to link to detailed information if desired.
- Provide a link to download a concatenated file of a series of Web pages so that users can print an entire document rather than printing multiple Web pages.
- Use backgrounds carefully; make sure that users can easily read the text of a page against the background.
- GIFs should be used for small graphics where there are a limited number of colors, and JPEGs should be used for photographic images.
- If a user links to another site, the site owner of the link must be informed; this will enable the site owner to notify everybody involved if the link changes.

However, users should consider the following facts:

- Too much graphics and animation slow down operations.
- Big pictures slow the loading of pages.
- Copyright of graphics should be granted.
- Proofreading is necessary.
- Browser compatibility must be checked.
- Avoid one-way streets in HTML documents.

4.4   CONTENT AUTHORING TOOLS

The first HTML tools focused on streamlining individual page creation, simplifying the need to memorize HTML page tags and syntax. They left site management up to those experts who were skilled with a whiteboard and with more sophisticated page layouts. Although hand coding HTML is still the preferred choice of many Web experts, developing and maintaining an entire Web site this way is very time consuming and not recommended. A Web site's foundation is proper management of all its files, directories, and links, and these seem to grow exponentially in any decent-sized site. Following this logic, the flip side to making it easier to stamp out Web pages is a tool that can manage and restructure the changing sites of the customers. As HTML authoring and site management applications mature, it is getting easier to find tools. Web site enhancement with pre–dynamic HTML (DHTML), cascading style sheets (CSSs), scripts, and pre–extensible markup language (XML) demands the use of powerful authoring tool environments.

As Web authoring becomes increasingly complex and the Internet diffuses into software, programs from word processors to databases all now export to HTML. This means the ability to both create and integrate HTML from other sources grows in importance. Similarly, tools recognizing that modern Web sites are the work of many people across different departments also get support. HTML was never designed to be a precise layout language, and even now, truly accurate page layout as with desktop publishing is the rare exception in Web environments. To sidestep this, designers have developed numerous workarounds and clever hacks. The most common position tricks are old single-pixel transparent GIFs and the extensive use of tables.

The creators of Netscape and Microsoft disagree on the best means of achieving absolute position via CSS, which means users have to wait before there is a single standard way to address this problem. Providing users with a means of simple HTML page production while controlling access to that production guarantees an easier life for Webmasters. For departmental users, an easy-to-use, page-focused HTML authoring package is the best choice. In the graphics area, the best tools can do on-the-fly, drag-and-drop importing and perform basic inline image touchups, such as resizing and resampling, without the need to launch a separate graphic editing application. These are must-have tools for professional-level development; similarly, any site management–enabled tool worth the download must verify and correct links in conjunction with directory and file moves, saving the time to recode pages.

Users should take advantage of free downloading of Web authoring tools for tests. Those tests can determine whether the authoring tool performs well in a particular environment.

Table 4.1 summarizes a couple of selection and test criteria for Web authoring tools (Santalesa, 1997).

There are many tools available for generic use. Examples include: PageMill from Adobe, Homesite from Allaire, HomePage from FileMaker, Dreamweaver from Macromedia, CyberStudio from Golive, Hotmetal from Softquad, FrontPage from Microsoft, Fusion from NetObjects, and VisualPage from Symantec. This list is not complete and is constantly changing.

TABLE 4.1
Selection Criteria for Web Authoring Tools

General features:
    Frames editor
    Table editor
    Forms editor
    HTML source editor
    Wizards/templates
    User-defined tags
    HTML source code editing
    HTML code validation
    Multiple undo
HTML support and importing:
    HTML version's support
    Font definition and support
    CSS
    External CSSs
    Dynamic HTML
    Import MS Word
    Import WordPerfect
    Import rich text format
    Import spreadsheet tables
Media and extensibility:
    Java applets
    JavaScript
    ActiveX controls
Graphics:
    PNG
    BMP
    Automatic file conversion
    Client-side image maps
    Server-side image maps
Site management:
    Link verification and correction
    Global search and replace
    Overall site items
    Modify site structure on the fly
Vendor:
    Number of clients
    Founded when
    What other products are offered
    Support (e.g., on site, hotline)
    Maintenance contracts
    Financial strength
    Keeps current with servers and their software releases

4.4.1   PAGEMILL FROM ADOBE

With this tool, there is no need to learn HTML or master complex applications. With its drag-and-drop simplicity, PageMill integrates seamlessly with current office and graphics applications, including Word, Corel, WordPerfect, Excel, Photoshop, Illustrator, and many others. After the site is built, delivery to the Web is easy using the built-in uploader. Integrated site management features enable users to keep the site up-to-date almost automatically, with advanced link management and site-wide search and replace capability. It offers the following features:

- Integrated site management tools enable efficient site maintenance.
- Font face support allows designer to specify text fonts.
- User interface enhancements streamline the work environment and save screen space to maximize the work areas.
- Support for Java and ActiveX allows users to preview Java applets and ActiveX controls within PageMill.
- Enhanced frame features support the use of borderless frames.
- Internet Explorer from Microsoft lets users preview pages as they appear in Internet Explorer and provides live Web browsing capabilities.

This product provides all the necessary tools to get a company's business on the Web quickly and easily, including an intuitive WYSIWYG interface, drag-and-drop page creation, Photoscope, Internet Explorer, integrated site management features with advanced search-and-replace capabilities, and more than 10,000 Web-ready images and animations.

PageMill's capabilities for building Web pages include:

- WYSIWYG interface: PageMill provides an intuitive interface so the user can create professional-looking Web pages quickly and easily. Access common functions and commands on the convenient button bar or via pull-down menus. At every step, see what visitors on the sites will see with PageMill's view-as-you-create interface or preview the pages in popular Web browsers.
- Drag-and-drop simplicity: Create Web pages by typing in text or dropping in content from current office and graphics applications. There is no need to type in long and cumbersome URLs or hyperlinks. It is also possible to add sophisticated media elements, such as animations, sounds, movies, and Java applets.
- Frames: Make information attractive and accessible by adding frames to Web pages. Simply click and drag to drop in text, images, and hyperlinks from the desktop. A frame border checkbox lets the user specify whether the frames have visible borders. The user can change and rearrange frame elements directly in the documents without wading through dialog boxes.
- Tables: Organize content using tables filled with text, images, videos, or even other tables. Also, spreadsheets can be cut and pasted. Enhanced table support allows for easy selection and better control over table and cell height, and cells can be easily joined or split, inserted or deleted.

Capabilities for enhancing pages include:

- Photoshop: This image-editing software creates high-quality images. It can be used to adjust scanned images, correct colors, and apply filters for special effects. Users can also create logos, buttons, icons, navigation controls, and background textures, all with drag-and-drop simplicity.
- Ready-to-use templates, images, animations, and sounds: PageMill includes thousands of Web-ready images, animations, Java applets, and customizable templates to help users create sites and bring them to life.

Site management capabilities include:

- Site overview: View all of site's resources at a glance in a Windows Explorer–type tree display. When files are renamed or rearranged, links are updated automatically.
- Graphical links view: A graphical view enables users to see a visual map of links and all other resources.
- Site-wide spell checking and search-and-replace capability for text, images, and links: These make site maintenance fast and efficient.
- Errors directory: PageMill automatically analyzes the sites and reports errors in an errors directory. These are easily repaired by dragging the correct file over the error, which automatically fixes incorrect references.
- Externals directory: Files from locations outside of their intended site folder are not included when a site is uploaded to a Web server, thus appearing as broken links. PageMill allows users to copy external local files into the resource folder for upload.

Advanced capabilities include:

- Built-in HTML editor: PageMill provides an integrated source code editor so users can add and format HTML directly. Users can quickly toggle between source code view and regular preview modes.
- Support for Java and ActiveX controls: Users can place and preview Java applets and ActiveX controls directly in PageMill pages.

To post sites, users can transfer them directly to the Web using the built-in FTP uploader, which will upload the entire site or just changes to any Web server, without requiring any special server extensions.

4.4.2 Homesite from Allaire

This product features an intuitive WYSIWYG interface and an extensive set of HTML and page management tools. Users are able to create pages quickly, maintain maximum control over HTML coding and execution, and preview pages in its built-in browser. The library of wizards simplifies leveraging emerging Web technologies. Principal features of the product include the following:

- Jump-start development efforts with prebuilt templates and reuse templates across projects by clicking for a snapshot of the template and wizards window.
- Open, toggle, and track edits among multiple documents by clicking for a snapshot of the document tabs.
- Open HTML documents directly from the Web with just a URL address.
- Access remote FTP servers for easy uploading and downloading of files by clicking for a snapshot of the FTP dialog.
- Browse images and image libraries directly within Homesite before dragging and dropping them to the particular page by clicking for a snapshot of previewing thumbnail images.
- Call on wizards to generate HTML tags and pages quickly.
- Display the tag editing dialog box for modifying attributes.
- Preview HTML pages in real time, including server-side scripts, via direct links to Microsoft Explorer from within Homesite.
- Update entire projects, folders, and files simultaneously and use regular expressions and wildcards with the find-and-replace capability.

High-quality, high-speed Web pages are supported by built-in page management tools. The attributes are:

- Track and troubleshoot code at a glance with automatic color coding of HTML and other client and server-side scripts (JavaScript, CFML).
- Correct content with an integrated spell checker that finds typing mistakes without highlighting HTML tags.
- Test links and modify documents and connections on the fly by clicking for a snapshot of link verification.
- Validate HTML syntax.
- Weight Web documents to check file sizes, including dependent files such as graphics, and calculate estimated download times at different modem speeds.

This product supports the latest Web technologies with the result that extended Web functionality can be built. Advanced features include:

- Use CSSs to take advantage of the powerful style sheet features in the latest browsers to define and save their own styles, preview styles, and apply them to existing text.
- Create and edit client-side scripts quickly and build in JavaScript functionality such as menus and scrolling text easily.
- Use the frames wizard to create multiframe documents and build complex, frame-based Web sites quickly.
- Call up specialized wizards for tables, DHTML, progressive networks RealAudio, and RealVideo to incorporate these technologies on the pages.
- Add tags and scripts from Cold Fusion (Allaire) and Active Server (Microsoft) page extensions.

Homesite can be configured and customized with shortcuts and personal preferences. It is recommended that users:

- Customize main and tag area toolbars for fast access to the custom tags and HTML features used most often
- Assign reusable code to custom toolbar items for easy access
- Edit tags easily and extend the tag set with user-defined tags and dialog boxes
- Add custom help files

4.4.3 HOMEPAGE FROM FILEMAKER

HomePage is a Web page authoring tool that helps users design and develop powerful, customized Web pages rapidly, without having to learn HTML. It delivers the capabilities in a cross-platform solution that provides ease of use as well as the flexibility and power experienced users want.

Use the FileMaker Connection Assistant to create interactive forms that connect directly to FileMaker Pro databases. The tools are:

- Form design elements: Create interactive forms easily using premade elements. Select from text field boxes, pop-up menus, check boxes, radio buttons, password fields, and submit and clear buttons.
- Frames: Divide a page into separate sections of varying sizes, each of which act independently of each other. Users can preview the frames while working and can quickly resize and position them.
- Tables: Create multirow and multicolumn tables instantly with a few mouse clicks, and specify colored backgrounds and cells. Tables are generated automatically when tabular text (e.g., spreadsheet information) is imported into home pages.
- Find and replace: Quickly search the entire site to find and replace any specified word or text block.
- Libraries: Store, organize, and preview frequently used images, movies, animations, text, or HTML code for easy use and reuse.

Advanced features include:

- Applets and scripts: Design interactive sites using Java applets and JavaScript.
- Multimedia support: Supports multimedia plug-in data files such as Quick Time movies for added site interactivity.
- Image maps: Easily add client- or server-side image maps to link specific sections of graphic images to other pages and sites.
- Document statistics: Displays estimated download times for the entire Web page or any part of the page being created.
- Color-coded HTML: Use color-coded HTML tags to help the user identify and edit headers, links, plug-ins, and more.
- Text wrap: Automatically wraps text around tables and images.

Automatic features include:

- Intuitive site management: Use the site editor to get a complete outline of all pages and elements for a particular site.
- Link verification: Identify and repair broken links within particular sites.
- Site upload: Upload sites remotely to a Web server without the need for additional FTP software. Automatic file consolidation helps prevent missing images, animations, or other content.
- Automatic links: Automatically create, copy and paste, or drag-and-drop links onto particular pages.
- Automatic graphics conversion: Automatically convert PICT or BMP images into GIF format when users drag/drop or paste them from other applications.
- Background preview: Display background images from within home pages created by FileMaker.
- Spell checking: Automatically check sites for misspelled words and typos.

Capabilities of the user interface include:

- Intuitive interface: Use familiar tools and actions to easily build and design Web pages. Easily place and format text, graphics, and links on Web pages.
- Assistants: Speed through creating sites with assistants that take the user step-by-step through the process — for doing everything from establishing FileMaker Pro database links to creating frames.
- Automatic programming: The home page does all the complicated HTML work behind the scenes, but the user still has the power to access the raw HTML code for advanced editing and customization of pages.
- Built-in clip art and templates: Select from an expanded library of over 2500 attractive clip art images and 45 complete site templates to build great-looking Web pages more quickly.

4.4.4 DREAMWEAVER FROM MACROMEDIA

Dreamweaver is the visual tool for professional Web site design. It offers roundtrip HTML between visual and source editing and provides support for creating cross-browser DHTML. It gives the productivity of a visual Web page layout tool with the control of an HTML source editor. It is a visual authoring tool that allows simultaneous visual and source HTML editing. JavaScript applications are also included. It manages content across an entire site rather than page-by-page.

Its principal capabilities are as follows:

- It can create pages for older browsers. In fact, it has a Browser Targeting feature that helps debug pages for the target browsers.
- Dreamweaver supports Shockwave and all plug-ins, but does not create Shockwave movies. Its native format is text. However, it is great for pulling all Shockwave content together in a Web site.

- It is actually an HTML editor. Although it can be used to create cool DHTML animations, it does not offer the same depths as Director or Flash for creating animations. However, it does offer depth in features for creating complex and dynamic Web pages.
- The main goal for Dreamweaver is to be a visual HTML tool adopted by Web professionals. Currently, some professionals use text editors such as BBEdit or Homesite. The goal is to work with these editors, not to replace them.
- Macromedia is committed to cross-platform solutions. DHTML represents the latest way to develop Web-based multimedia content for both the Macintosh and Windows platforms. DHTML works in both Netscape and Internet Explorer. Since those browsers are now widely accepted, DHTML provides a good solution for simple multimedia without using a plug-in.
- DHTML should be used in combination with plug-ins. Plug-ins like Shockwave Flash and Director offer complex interactivity and rich multimedia experiences that can be difficult or impossible to achieve with DHTML, and they are compatible with older browsers. DHTML offers an interactive alternative to animated GIFs, and it can be used on its own or in conjunction with plug-ins and ActiveX controls.

4.4.5   CyberStudio from GoLive

CyberStudio is a tool for creative control and flexibility. Its ease of use makes it a popular tool for content authoring. Its principal features will be summarized.

Visual layout and design control features allow the user to:

- Precisely place objects on pages, using a layout grid and boxes
- Play multimedia and Java applets while working in a WYSIWYG mode
- Visually create and lay out tables
- Visually create and edit Web pages with frames
- Create Web-based forms
- Easily create clickable image maps
- Drag ActiveX objects onto individual pages
- Place information directly in the HTML page header
- Select from a wide range of color palettes, including Web-safe colors
- Simultaneously support multiple language character sets

Advanced site management features help the user to

- Design, create, import, edit, and view an entire site
- Visually check and repair broken links
- Spell check a single page or a complete site
- Easily download or upload individual Web pages or an entire Web site via FTP
- Quickly open and edit HTML documents on Web servers

- Maintain sites with pages on multiple Web servers
- Display the estimated download times for a Web page

Features related to HTML native file format include:

- Always works in an HTML native file format
- Allows user to share files with others — using any other HTML authoring tool or text editor — on any type of computer
- Allows user to read, write, and edit unrecognized tags.

HTML source code and JavaScript editing capabilities allow the user to:

- View and write HTML source code
- Write HTML code snippets
- Automatically check HTML syntax
- Access a fully editable HTML tag database
- View the HTML tag hierarchy of a Web page in a clear-cut outline view
- Create JavaScript using the integrated JavaScript editor with color syntax checking
- Change page elements and styles individually or throughout an entire Web site using find and replace.

User-defined options include the ability to:

- Save user-defined elements on a palette then easily drag and drop them onto Web pages
- Save user-defined colors within a project file
- Save user-defined font sets within a project file
- Save and easily access HTML pages as stationary files

4.4.6 HOTMETAL FROM SOFTQUAD

This product is for both professionals and people new to Web site development. Both simple and sophisticated Web sites can be developed with this authoring tool. Product highlights are:

- Enhanced productivity with three editing views
- Integrated authoring and site management tools in one application
- Comprehensive site management
- Easy site development for new users
- Support for all the latest Web technologies
- Quick access to Web assets with the Resource Manager
- Valid HTML that speaks to any browser
- Customizable user interface
- Tools to create great looking graphics
- Advanced accessibility support

Capabilities of the product include:

- WYSIWYG editing: This includes accurate displays, improved table editing, CSS display, and integrated browser preview.
- HTML source editing: Full-featured source code mode for total control over HTML, optional line numbering with dynamic line wrapping, auto formatting and color coding, and automatic tag completion are included.
- Tags on editing: Unique view combines the precision of source editing with the convenience of WYSIWYG, tags on WYSIWYG table editing, and tags tips to let the user instantly view all attributes set on a tag.
- Attribute inspector: The attribute inspector completes access to all element attributes; context-sensitive display shows valid element attributes; includes drag-and-drop text, effects, URLs, and more.
- Site maker: It builds an entire linked site; chooses from business, intranets, or personal Web sites; and selects from a wide variety of layouts and decors.
- Graphics creator: It includes PhotoImpact; easily adds graphics; creates drop shadows, buttons, textures, and more; and creates animated GIFs with Ulead's GIF Animator.
- Resource manager: It manages and uses images, site files, code, and other components; creates and manages asset pages with drag-and-drop ease; and offers a unique view that combines the precision of source editing with the convenience of WYSIWYG. Prepackaged effects include graphics, backgrounds, templates, Java applets, and DHTML. It adds components to existing categories or creates new ones.
- CSS support: It supports external style sheets, style elements, and style attributes, and WYSIWYG displays in the editing environment.
- Support for latest Web techniques: It supports the following techniques: CSS, Frames, DHTML, Java, JavaScript, VBScript, Miva, XML, WebTV, ActiveX, Shockwave, Flash, Quicktime, RealAudio, streaming audio and video, and many Macros.
- Database import wizard: It creates queries without knowing structured query language (SQL); supports MS Access, Excel, and other open database connectivity (ODBC) sources; and imports and converts files from Word, WordPerfect, AMIPRO, RTF, ASCII, and many more.
- Frame editor: It creates, edits, and imports sophisticated frames with drag-and-drop ease, it offers pixel-perfect frame editing, and supports one-button creation of borderless frame sets.
- Valid HTML generator: It writes always 100% valid HTML; it checks rules to ensure clean HTML; and the extensible rules file allows new tag definitions.
- Integrated site management tools: It offers site-wide visualization in WebView; it easily finds and fixes broken links; site summary pages provide site statistics such as file sizes and download times; it has automatic links updates, when moving files; publishing to multiple servers; site-wide searches and replacements of text; HTML tags and attributes are

supported; one-button publishing is supported for an entire site or just for changed pages.

- Improved interface: It offers an enhanced drag-and-drop functionality, customizable workspaces and toolbars, and workbook mode feature tabs for open documents.
- Accessibility support: It supports visual dynamic keyboards and screen readers, validates pages for accessibility, and offers authoring assistance to ensure accessibility.

4.4.7 FRONTPAGE FROM MICROSOFT

FrontPage is an effective way to create and manage professional-quality Internet or intranet sites without programming. It makes it easy for new users and professional Web developers alike to build and maintain great looking Web sites rapidly.

Capabilities and features of the product include:

- FrontPage themes: More than 50 professionally designed themes (or graphical designs) provide users with consistent backgrounds, bullets, banners, hyperlinks, and navigation bars across the entire Web site — all without hiring a graphic designer.
- The navigation view: It lets users create and manage the navigational structure of Web sites within a short period of time. It is easy to build and connect new pages and move pages around. Users can even print a map of the entire site.
- Shared borders: Shared borders allow users to specify shared headers and footers or right and left margins across the pages. This provides a great deal of design flexibility in creating attractive yet consistently designed Web sites.
- Automatic navigation bars: Easily add universal navigation bars to the Web based on the site's navigational structure. If the structure is modified, FrontPage will update the navigation bars automatically, saving time and keeping all the links current.
- WYSIWYG table editing tools: The new editing tools and the mouse are expected to be used to draw and erase entire tables, rows, and columns on Web sites. The drag-and-drop feature is used to resize table cells and move or copy table rows and columns.
- WYSIWYG frame page editing: Create and view WYSIWYG frame pages and edit them directly on the screen in the FrontPage editor. Simply drag a frame border to add a new frame on the page or to change the size of an existing frame, or easily drag and drop content between frames.
- Form save results to e-mail: With the new form save results FrontPage component, users can easily create a form that will send submissions directly to designated e-mail addresses. What used to require a custom script requires no programming in FrontPage.
- Hover buttons: FrontPage enables users to automatically create small Java applet hover buttons, so that when a user "hovers" over the buttons or

clicks on them, they will change colors, change shape, or animate any way the user has chosen.

- Banner ad manager: Automatically create rotating banner advertisements on Web pages. Users just specify the banner images they would like rotated and choose the transitional effect between the images. It is easy to add rich content to existing pages.
- Integration with Internet Explorer: Web pages can be edited in FrontPage; changes can be saved back to the Web server.
- Channel definition format (CDF) wizard: With the CDF wizard, users can quickly turn their Web site into a channel that users can subscribe to — for automatic delivery of their Web content to any desktop running Internet Explorer.
- DHTML support: FrontPage allows users to easily add DHTML features to Web sites without programming. Simply use the FrontPage menus and dialogs to add text animations, collapsible outlines, page transitions, and more to individual sites.
- CSS support: FrontPage makes it easier to create and design impressive Web pages through support of CSS. CSS allows users to define complex styles for titles, paragraphs, headers, and more.
- FrontPage image editing tools: Editing tools — which are improved continuously — make it easy for users to bevel, crop, flip, rotate, or automatically wash out images on Web pages. FrontPage also lets users shrink or resample an image in order to reduce its download time.
- Editor views tabs: New editor tabs let users quickly toggle between views of Web pages. The normal tab is for WYSIWYG editing, the HTML tab is for editing HTML directly, and the preview tab lets users view Web pages in browser mode from within the FrontPage editor.

Advanced capabilities and features of the product include:

- Easier hyperlinking: A new hyperlink dialog box simplifies the process of linking to new pages, existing pages in a Web, pages on the Internet, e-mail addresses, bookmarks on a specific page, or targets within frame sets.
- One-button Web publishing: FrontPage makes publishing sites to the Web easier than ever by publishing only those files that have changed since last published. FrontPage will automatically detect changes made by others in a multiuser environment.
- Improved import wizard: FrontPage now makes it easier than ever to import existing Web sites and existing content from a computer's file directory structures, a Web server, or a URL location on the WWW.
- Simplified form save results: Setting how the Web handles the results of forms is now easy with improved form save results. An intuitive and streamlined user interface gives more power and control so users can set up how specific forms are handled.

The benefits of the product can be summarized as follows:

- Its easy-to-use, leading edge features let users create professional Web sites without programming. Create WYSIWYG frame pages and draw HTML tables in the WYSIWYG FrontPage editor. Drop in sophisticated, interactive functionality using FrontPage components.
- The comprehensive management tools let users quickly build and maintain well-organized Web sites. With automatic hyperlink maintenance, users never have to worry about broken links. Plus, flexible collaboration features let users work with others on various Web sites.
- Seamless integration with existing content and with desktop applications users already have makes users productive from the start. In addition, strong browser integration makes it easy to customize and view Web site content.

4.4.8 Fusion from NetObjects

Fusion is a powerful Web site creation and management tool that addresses design and maintenance issues. Most Internet-based standards are supported. Web sites are continuously changing. They are different today than they were even a year ago, and so are the persons who build them. As business requirements have increased, sites have become larger and more visually enticing; they are alive with animation and special effects; and they need to work on all browsers. Visitors also expect sites to be interactive. They want to complete transactions and conduct business from their desktops. They want real-time data on demand. Until now, creating Web sites meant working with the limitations of products designed for obsolete sites: page-oriented, code-based tools or visual tools that offer limited HTML control. Net-Objects' Fusion helps to build the dynamic Web sites its clients require. It delivers an open site-building environment with flexibility and layout control as well as robust site management and cross-browser support. Using a site-oriented approach, Net-Objects adds more than 150 features and enhancements that let the new breed of Web professionals create sites the way they want, with no limitations.

Product features include:

- Open architecture: Enjoy broad support for industry technologies and editing tools.
- SiteStructure editor: Create order out of chaos — whether the user has mountains or molehills of information.
- Three-way layout editor: Create effective pages using any combination of three new modes for maximum control and flexibility.
- DHTML actions: Actions speak louder than words; it easily creates DHTML-based animation and interactive effects.
- Everywhere HTML: Don't let a browser get in the way of users. Build once, publish anywhere.
- Total publishing control: Organizes content the way users want. Choose from three predefined directory structures, or create an individual structure.

- HTML coding and scripting: Either the user or Fusion creates the HTML code.
- Database publishing: Link to external ISAM (index sequential access method)- or ODBC-compatible databases or use components developed by partners from NetObjects for dynamic database publishing on the e-business site of the user.
- SiteStyles editor: Globally applies a visual theme to Web sites by using or customizing standard styles.
- Fusion components: It creates interactive features and turns visitors into customers.

4.4.9 VISUALPAGE FROM SYMANTEC

A word processor style user interface with toolbars and extensive drag-and-drop support makes creating professional looking Web pages an easy exercise. The intuitive user interface of the product also results in a greatly reduced learning curve so the user can produce results within a short time.

VisualPage provides users with what they actually need to make all new Web sites fast. In addition, it can help give existing Web sites a swift, successful makeover. VisualPage automatically gathers the information it needs to help manage Web sites. Links to files are instantly checked and changes are automatically updated across the entire site. Users can ensure professional text presentation with site-wide spell checking and site-wide search and replace. Furthermore, the entire Web site can be updated on the Web server with a click of a button. As a result, productivity is greatly improved as users maintain and enhance Web sites.

Capabilities and features of the product include:

- Cascading style sheets: WYSIWYG support for creating and using CSSs makes it easy to create a uniform look and feel across the entire Web site. Style sheets allow users to separate content of Web sites from style. Using them, users can add content to Web sites and then apply a style to see how it looks. If users don't like the look, they simply change the style and the whole Web site is instantly updated with the new look. Users don't need to manually update every element in the Web site. The result is an increase in the speed with which users can create Web sites and an increase in the probability that the Web site will have a uniform look and feel. A set of sample style sheets is included with the product.
- Real-time WYSIWYG: VirtualPage represents the Web page the same way it appears in the viewer's Web browser. It also instantly reflects how changes made in the visual editor will appear in the Web browser. For example, as users resize a table cell by dragging its border, the text in the cell automatically adjusts itself to fit the new cell size. (Other tools wait for the border to be dropped before they show the results of the resizing.) The result is that users can make the Web page look just right, in less time.
- Layers and absolute pixel positioning: The WYSIWYG support for layers allows users to place objects anywhere on the Web page with absolute

pixel positioning. Layers allow users to design Web pages using a desktop publishing layout scheme as opposed to a word processing layout scheme. This allows users to position objects anywhere on the Web page without affecting the positioning of other objects. With VisualPage, users can easily use desktop publishing and word processing layout schemes within the same Web page. The result is complete control over the Web page layout.

- Easy table and frame creation: The product makes creating and customizing tables and frames very easy. Create a set of frames or insert a table with the click of a button. Resizing the table's cells is as easy as dragging the border to reflect the new size. The result is a professional look and feel for Web pages.

- Font and font group support: VisualPage allows users to visually apply fonts and font groups and to create customized font groups. Font groups are used to specify alternative fonts in case the platform on which the customer is viewing the particular Web site does not contain the user's first choice.

- Web page preview and test environment: The product includes a Web page preview and test environment where users can click on links to follow the flow of the links and can see Java applets execute. Users can also easily load a Web page into any external browser for final testing. The result has the user's intended look and feel.

- Open industry standard HTML: VisualPage generates well-formatted, industry-standard HTML files that are easy to understand and modify. It does not generate any proprietary code that limits where Web pages can be hosted. The result is peace of mind that the Web pages will work with all customers' Web servers and Web browsers.

- Color-coded syntax editor: The product provides a color-coded syntax editor so users can work at the code level if they wish. The syntax editor represents the HTML in different colors to differentiate tags, text, attributes, values, and comments. The HTML is also automatically formatted in a way that makes it easy to read and understand the code. Any changes made in the syntax editor are instantly reflected in the visual editor and vice versa. The result is that users can create Web pages using the environment that feels most comfortable to them.

- Visual page art collection: This is a collection of over 12,000 professionally designed backgrounds, banners, page dividers, and color-matched sets. These images are royalty free and will help users quickly build a unique and professional looking Web site.

- Web page templates and samples: This is a set of professionally designed Web page templates and samples that can be used to quickly create professional looking Web sites. Use these templates in conjunction with the style sheet templates.

Advanced capabilities of the product include:

- AutoImport: After specification of home pages, VisualPage quickly creates a project file containing information about the files that make up the Web site. The project file allows users to quickly understand the Web site structure, to automatically maintain the integrity of the site, and to greatly improve the speed with which they can update and enhance the site.

- Project manager: After Web site import or creation, VisualPage's project manager provides four distinct views of the Web site:

  File view — Enables users to easily manage Web sites using a browsable view of the local Web site directories and Web sites.

  Server view — Provides users with a complete understanding of how particular Web sites appear on Web servers.

  Link view — Delivers a graphical view of the Web site, allowing users to quickly grasp the site's structure.

  Status view — Presents a list of the links that are invalid or missing, giving users the means to quickly establish site integrity.

- Automatic link checking: Web site links are automatically checked in the background while users build their Web sites. Invalid and missing links are highlighted in both the link and status views. Invalid links are links to files that are located outside the local Web site's directories. These links will be broken when Web sites are published. Missing links are links to files that don't exist. These are caused by files that are moved or removed after links are made to them. Fixing broken links from Web sites is as easy as selecting the broken link and browsing to the file that should be linked. Once the link has been fixed, all files that contain the same broken link are also automatically fixed.

- Automatic link updating: Once a project file is created, all links are automatically maintained for the users while they build and modify their Web sites. Changing a file name results in all links to that file automatically being updated to the new name. Changing an e-mail link in any of the project manager views results in all Web pages that contain the e-mail link being updated.

- Site-wide search and replace: Updating content across the entire Web site is easily achieved using the site-wide-search and site-wide-replace features. No more worrying that somebody has forgotten to check all Web site files for content that should have been changed.

- Site-wide spell checking: Site-wide spell checking helps to ensure that users avoid embarrassing spelling mistakes, which can detract from a professional image.

4.5   SITE PERFORMANCE OPTIMIZATION PRODUCTS

The products introduced so far are capable of supporting content authoring and deployment. Some built-in features help to create optimized content. Usually, however, these features do not check the performance of pages when under stress, caused by many visitors hitting the same pages simultaneously. Site performance optimization tools address just this. They not only check for incompleteness and errors, but also try to repair problems.

It would be optimal to test how Web applications will perform on public and private Web sites before they are deployed. Such testing applications are supposed to help in the following areas:

- Testing how sites will react to numerous end user hits
- Monitor application interaction
- Measure Web access performance from the end user's perspective
- Correlate events with load profiles
- Analyze impacts of traffic load on selected clients trying to access a Web site
- Predict the performance of Web servers under stress
- Help to optimize Web servers and bandwidth for traffic profiles

WebLoad from RadView is an excellent example for a preventive optimization of site performance. Other examples are from Mercury Interactive, Rational Software, and Segue Software.

This section focuses on site management tools only. It gives practical examples with two products.

4.5.1 LINKBOT FROM TETRANET

Linkbot is a suite of Web site management utilities that can help track down and repair problems on Web sites. It contains all the tools Webmasters need to automate site management and maintain an error-free site. The principal features are:

Basic features	Description
Finds and repairs dead links	Improves the quality of sites by cleaning up bugs
Finds unused orphan files	Recovers disk space by cleaning server of files that are no longer linked
Finds stale content	Identifies pages that have not been updated recently and may not be accurate anymore
Finds slow pages	Ensures that users with slow connections can access pages quickly; addresses the problem of large pages that are slow to download
Finds pages with missing titles	Pages without titles will not display correctly in browsers and search engine query results; finds and repairs pages with missing titles
Supports workgroups	Produces separate reports for subsections of the site belonging to a specific author
Interactive site mapping	Explores the structure and organization of sites in Linkbot's Explorer-style interface
Multitask scanning	Scans large sites quickly by processing up to 30 URLs at a time
Automated scheduling and HTML reports	Schedules Linkbot to scan the site at regular intervals and to automatically generate HTML reports detailing the site's major problem areas

Checks internal and external HTTPs and FTP links — Isolates all broken links on a site

Checks syntax of "mail to" URLs — Finds "mail to" references that may be invalid

Filters URLs from the analysis — Specifies areas of the site that Linkbot should not check

Mapping and organization — Description

Maps out the structure of the site in an "Explorer"-style interface — Allows Webmasters to interactively explore the organization of a site in a familiar interface

Shows links in and out of a selected URL — Makes editing bad links easier by showing all the links that point to a selected URL

Sorts the site's URL list by title, description, author, size, last modified date, and type — Isolates pages with missing titles, finds all the documents published by a specific author, finds old and new files, and finds large files with slow download times

Provides advanced filters for viewing subsets of a site's contents — Makes it possible to view and organize smaller subsets that contain thousands of URLs

Maintenance and repair — Description

Provides built-in HTML editing — Edits and repairs problems within Linkbot's interface

Rechecks specific subsets of URLs on a site — Enables Linkbot to check targeted groups of URLs (e.g., all broken links to external sites)

Allows Webmasters to export results to a delimited file — Allows the results of Linkbot's analysis to be exported to a file and then imported into a database

HTML reports — Description

Creates nine customizable HTML reports — Summarizes the site's problem areas and statistics; automatically generates detailed reports on broken links, broken pages, warnings, orphaned URLs, and on new, old, and slow pages

Workgroup features — Description

Creates reports listing the site's problems by author — Makes it easier for authors in a workgroup to repair the problems in their pages

Automation/scheduling — Description

Automates maintenance with customizable scheduling options — Allows Linkbot to be programmed to turn itself on, check a site, and produce a report

TABLE 4.2
What Is Slow Report by Tetranet

PAGE	SIZE (kb)	28.8	ISDN	
http://www.tetranetsoftware.com/products/linkbot-reviews-index.htm	134	00:00:46	00:00:20	00:
http://www.tetranetsoftware.com/products/linkbot-reviews.htm	113	00:00:39	00:00:17	00:
http://www.tetranetsoftware.com/partners/index.htm	96	00:00:33	00:00:15	00:
http://www.tetranetsoftware.com/products/linkbot-tour.htm	94	00:00:32	00:00:14	00:
http://www.tetranetsoftware.com/index.htm	87	00:00:30	00:00:13	00:
http://www.tetranetsoftware.com/products/linkbot.htm	80	00:00:27	00:00:12	00:
http://www.tetranetsoftware.com/products/wisebot.htm	78	00:00:27	00:00:12	00:
http://www.tetranetsoftware.com/products/index.htm	77	00:00:26	00:00:12	00:
http://www.tetranetsoftware.com/partners/text.htm	74	00:00:26	00:00:11	00:
http://www.tetranetsoftware.com/products/wisebot-reviews-index.htm	68	00:00:23	00:00:10	00:
http://www.tetranetsoftware.com/products/wisebot-tour.htm	65	00:00:22	00:00:10	00:
http://www.tetranetsoftware.com/products/linkbot-main.htm	59	00:00:20	00:00:09	00:
http://www.tetranetsoftware.com/buynow/index.htm	59	00:00:20	00:00:09	00:
http://www.tetranetsoftware.com/products/wisebot-main.htm	57	00:00:20	00:00:09	00:
http://www.tetranetsoftware.com/products/text.htm	56	00:00:19	00:00:08	00:

Linkbot offers a number of reports regarding performance evaluation of pages and links. Table 4.2 displays download times using modems, ISDN, and T1 links. The summary report (Table 4.3) offers URL and site statistics. The warnings report (Table 4.4) concentrates on errors, temporary redirects, and miscellaneous warnings. The problem pages report (Table 4.5) displays broken pages. This list of pages may be organized by authors. The broken link report (Table 4.6) summarizes and details link-related errors.

4.5.2 SITESWEEPER FROM SITE/TECHNOLOGIES

Web professionals are challenged to maintain the integrity of mission-critical Web-based business applications. SiteSweeper performs time-consuming quality assurance tasks for the Webmaster and provides the additional information needed to ensure the smooth operation of any Web site.

Large images with slow download times are a major source of frustration for Web site users. SiteSweeper helps make sites more efficient by automating the process of determining the total download sizes of all the pages on the site. It even estimates the download times at different connection speeds for easy analysis and comparison. Time-intensive searches for bad links are also automated with

TABLE 4.3
Summary Report by Tetranet

URL STATISTICS

	HTML	IMAGE	MAIL	FTP	APPLET	OTHER	ALL
GOOD	92	112	12	0	2	40	258
BAD	3	2	1	1	0	2	9
WARN	0	0	0	0	0	0	0
UNKNOWN	0	0	0	0	0	0	0
TOTAL	95	114	13	1	2	42	267

SITE STATISTICS

	HTML	IMAGE	MAIL	FTP	APPLET	OTHER	ALL
INTERNAL	79	114	12	0	2	2	209
EXTERNAL	15	0	1	1	0	40	57
NEW	78	0	0	0	0	0	78
OLD	0	0	0	0	0	0	0
SLOW	15	0	0	0	0	0	15
ORPHANED	0	0	0	0	0	0	0

TABLE 4.4
Warnings Report by Tetranet

SUMMARY OF ERRORS

	HTML	IMAGE	MAIL	FTP	APPLET	OTHER	ALL
PERMANENT REDIRECTS	0	0	0	0	0	0	0
TEMPORARY REDIRECTS	0	0	0	0	0	0	0
OTHER	0	0	0	0	0	0	0
TOTAL	0	0	0	0	0	0	0

PERMANENT REDIRECTS: URL NOW POINTS TO ANOTHER LOCATION

None Found

TEMPORARY REDIRECTS: URL HAS BEEN TEMPORARILY REDIRECTED

None Found

OTHER WARNINGS: MISCELLANEOUS WARNINGS

None Found

TABLE 4.5
Problem Pages Report by Tetranet

PAGES WITH ERRORS BROKEN DOWN BY AUTHOR

AUTHOR	PAGES WITH ERRORS
(No Author specified in META tags)	3
rian Sharpe	1
TOTAL	4

BROKEN PAGES CREATED BY: (No Author specified in META tags)

PAGE: Starting URL http://www.tetranetsoftware.com/demo.html
BAD LINKS

1. http://www.tetranetsoftware.com/demo.html#broken (Bookmark that's broken, Error: Undefined Anchor)

2. http://www.tetranetsoftware.com/404.htm (404 Errors, Error: 404: Not Found)

3. http://hostnotfound.com/ (Host not found errors, Error: Could not connect)

4. ftp://ftp.tetranetsoftware.com/pub/broken.exe (FTP link, Error: Not found)

5. mailto:%20bademailaddress (Email with bad syntax, Error: Bad mail address syntax)

6. http://www.tetranetsoftware.com/linkbot35.gif (linkbot35.gif (4350 bytes), Error: 404: Not Found)

PAGE: Linkbot and Wisebot Downloads
http://www.tetranetsoftware.com/download/form-response.htm
BAD LINKS

1. http://tucows.matrix.com.br/files2/linkbot.exe (Brazil, Error: Could not connect)

PAGE: Linkbot Upgrades http://www.tetranetsoftware.com/support/upgrades.htm
BAD LINKS

1. http://www.tetranetsoftware.com/support/line.gif (line.gif, Error: 404: Not Found)

BROKEN PAGES CREATED BY: rian Sharpe

PAGE: Linkbot Download Page
http://www.tetranetsoftware.com/download/wisebot-download.htm
BAD LINKS

1. http://tucows.matrix.com.br/files2/wisebot.exe (Brazil, Error: Could not connect)

SiteSweeper, helping Webmasters identify and resolve problem areas before they escalate. Types of bad links — "not found," "unauthorized," and "moved permanently" — are identified and described in the reports of the product.

Informed decisions may be made about the state of the Web site. The following data are available:

- Properties of each page (Table 4.7)
- Aggregate site statistics
- Link status (Table 4.8)
- Links to external sites

TABLE 4.6
Broken Links Report by Tetranet

♭ SUMMARY OF ERRORS

	HTML	IMAGE	MAIL	FTP	APPLET	OTHER	ALL
Not Found	1	2	0	1	0	0	4
Can't Connect	1	0	0	0	0	2	3
Host Not Found	0	0	0	0	0	0	0
Time Out	0	0	0	0	0	0	0
Undefined Anchor	1	0	0	0	0	0	1
Other	0	0	1	0	0	0	1
TOTAL	3	2	1	1	0	2	9

♭ FILE NOT FOUND ERRORS - FILE DELETED OR REMOVED

LINK: 404 Errors http://www.tetranetsoftware.com/404.htm (Error: 404: Not Found)
PARENT LINKS
🔁 1. http://www.tetranetsoftware.com/demo.html

LINK: FTP link ftp://ftp.tetranetsoftware.com/pub/broken.exe (Error: Not found)
PARENT LINKS
🔁 1. http://www.tetranetsoftware.com/demo.html

LINK: linkbot35.gif (4350 bytes) http://www.tetranetsoftware.com/linkbot35.gif (Error: 404: Not Found)
PARENT LINKS
🔁 1. http://www.tetranetsoftware.com/demo.html

LINK: line.gif http://www.tetranetsoftware.com/support/line.gif (Error: 404: Not Found)
PARENT LINKS
🔁 1. http://www.tetranetsoftware.com/support/upgrades.htm

♭ CAN'T CONNECT ERRORS - SERVER EXISTS BUT NOT RESPONDING

LINK: Host not found errors http://hostnotfound.com/ (Error: Could not connect)
PARENT LINKS
🔁 1. http://www.tetranetsoftware.com/demo.html

LINK: Brazil http://tucows.matrix.com.br/files2/linkbot.exe (Error: Could not connect)
PARENT LINKS
🔁 1. http://www.tetranetsoftware.com/download/form-response.htm

- Catalog of images
- Pages with bad links
- Total download size of each page
- Thumbnails and size of all images

To use this product, the user simply enters the home page into the interface. Then SiteSweeper sweeps the site following each of the links identified. Along the way, it visits all the linked pages in the particular site and tests the validity of all links to

TABLE 4.6 (continued)
Broken Links Report by Tetranet

LINK: **Brazil** http://tucows.matrix.com.br/files2/wisebot.exe (Error: Could not connect)
PARENT LINKS
   1. http://www.tetranetsoftware.com/download/wisebot-download.htm

   HOST NOT FOUND - DNS LOOKUP FOR SERVER FAILED

None Found

   TIMEOUT ERRORS - SERVER FAILED TO RETURN THE FILE WITHIN THE SPECIFIE

None Found

   UNDEFINED ANCHOR - REFERENCED ANCHORS THAT WERE NOT DECLARED

LINK: **Bookmark that's broken** http://www.tetranetsoftware.com/demo.html#broken (Error:
Undefined Anchor)
PARENT LINKS
   1. http://www.tetranetsoftware.com/demo.html

   OTHER ERRORS - MISCELLANEOUS PROBLEMS

LINK: **Email with bad syntax** mailto:%20bademailaddress (Error: Bad mail address syntax)
PARENT LINKS
   1. http://www.tetranetsoftware.com/demo.html

TABLE 4.7
Properties of Pages by SiteSweeper

	Minimum	Average	Maximum
Embedded Components per Page	1	7	11
Links From per Page	2	6	9
Links To per Page	1	4	8
Page Download Size	3.4K	37.0K	54.4K
Age (in days)	0	21	25

TABLE 4.8
Link Status Report by SiteSweeper

	Pages	Internal	External	Total
OK	6	65	2	67
Unsupported Protocol	4	1	4	5
Anchor Name Not Found	1	1	0	1
Operation Cancelled	1	0	1	1
Not Found	1	1	0	1
Moved Temporarily	1	0	1	1

external Web sites. It gathers important data about each page it visits, including size, last modified date, and expiration date. When finished, it generates a series of reports, as specified by the Webmaster via the reporting interface. Since the reports are generated in HTML, they can easily be published and shared by the Webmaster.

Key features of the product are:

- Detailed reports on demand or by schedule
- Support for multiple servers
- Authentication for access to password-protected sites
- Parallel processing for fast sweeping
- Proxy server support

SiteSweeper is an industrial-strength application suitable for corporate Internet and intranet sites of any size. Multiple Web servers running on multiple platforms can be swept simultaneously.

Each sweep automatically generates a complete set of quality assurance reports in HTML format. The reports provide critical information to ensure the smooth operation of Web sites. Reports can be viewed and printed from universal browsers. All reports link to additional detailed information. To access this hyperlinked information, users are expected to click on the graphics and text hyperlinks within each report.

Typical reports are:

- Quality summary — This report gives brief summaries for items such as broken links, slow pages, missing ALT attributes, missing width or height attributes, distorted images, problem titles, and missing META tags.

- Problem pages — This report is like a "to-do" list that helps identify problems that need to be fixed. It lists all problem pages and types of problems. This report provides hyperlinks to the pages so Webmasters can see the problems immediately and directly. It also provides hyperlinks to all URLs and titles that were swept.
- Site atlas — SiteSweeper provides a site atlas that lists site configuration with all folders and subfolders and gives details on all resources used by particular Web sites. It also provides a resource utilization pie chart that shows the usage of text and images (GIF and JPEG) for Web sites.

SiteSweeper is a useful tool to conduct performance evaluation tasks with Web sites.

4.6 PERFORMANCE OPTIMIZATION AS A SERVICE

Performance optimization of Web applications may be offered as a service. Web-Garage is doing exactly this. For a fee, which is structured for various services, principal metrics of Web page performance can be analyzed and improved. All tuning activities are supported remotely. Depending on the service requested, the service provider determines the time it needs to fulfill the service.

The basic service, Tune Up, consists of the following diagnostics:

- Browser compatibility check — This indicates how well the Web page is displayed by viewing it with different browsers.
- Readiness check — This investigates whether the Web page is set up to be indexed correctly by search engines and directories.
- Load time check — This measures how long it takes for the site to load with an average-speed modem.
- Dead link check — This detects hard to find dead links on the customer's page.
- Link popularity check — This finds out how many Web sites are linking to a particular Web site.
- Spell check — This discovers and highlights spelling mistakes.
- HTML design check — This investigates how the site's HTML design compares to the best in practice.

Tune Up Plus offers more. Users work hard to get traffic to their Web sites, so Tune Up Plus ensures an efficient way of monitoring the quality of Web sites. This service is offered for more Web pages than the basic service. The most comprehensive service is Turbocharge. Its service components consist of Register-It, reregistration, browser snapshot, and GIF Lube.

The Register-It service registers the Web site with the top 100 search engines and directories on the Internet. This service includes:

- Automatic registration
- Selective registration (only selected directories are included)
- Free registration
- Free updates
- Status tracking

Whereas most services provide only one-time registration, this service provides reregistration and updates at no cost to the user. This company updates the top search engines list constantly and keeps abreast of the best search tools on the Internet.

The browser snapshot provides a snapshot of how the particular site actually looks on different versions of major browsers, platforms, and screen sizes. This service helps to verify that the particular site is compatible with all the various browsers on the Internet. This service includes:

- 18 screenshots in Netscape and Microsoft Internet Explorer on both Macintosh and Windows platforms
- Yearly license allowing multiple screenshots over time
- Free updates, including updates to the system, which covers support for additional browsers, platforms, and screen sizes

It is estimated that 30% of the sites on the Internet may suffer from browser compatibility issues. Getting visitors to a particular site can be expensive and time consuming. By using this service, customers can verify that users viewing the particular site from the various browsers see an accurate and complete picture. Also, the development of the site can be observed with this service.

Many search engines and directories use META tags to index Web sites. Users can greatly increase the odds that search engines will put the particular site near the top of the category by using the right META tags in the HTML code. META tags consist of two items:

- Keywords — These should match the words that someone would enter in a search to find a particular site. They generally include the following: company name, product name(s), product category, and the plurals and possible misspellings of these words.
- Description — This is the description that will be displayed with a particular listing in the search engine. It should make the user want to visit the site or solve a problem for the visitor.

GIF Lube helps pages load faster by reducing the size of an image. It reduces image size by reducing the number of colors in the image. GIF Lube also allows users to compress and convert images into GIF or JPEG formats.

A few reporting examples are demonstrated on a concrete site. Table 4.9 shows the list of compatibility warnings by browser types. The page summary is shown in Table 4.10. Finally, Table 4.11 shows page download times using various technologies, such as dial-in modems, ISDN, and T1.

TABLE 4.9
Compatibility Warnings by WebGarage

Compatibility Warnings by Browser:

Browser	Warnings
Netscape Navigator 4.0	0
Netscape Navigator 3.0	0
Netscape Navigator 2.0	2
Microsoft Internet Explorer 4.0	0
Microsoft Internet Explorer 3.0	0
America Online 3.0	0

TABLE 4.10
Page Statistics by WebGarage

PAGE SUMMARY	Fair
Browser Compatibility	Excellent
!Register-It! Readiness	Good
Load Time	Fair
Dead Links	Excellent
Link Popularity	Excellent
Spelling	Good
HTML Design	Poor

RECOMMENDATIONS

TABLE 4.11
Page Download Statistics by WebGarage

18 Total Objects on the Site
Total Size: 60138 bytes
Total Connects: 5

Connect Rate	Connect Time
14.4K	41.99 seconds
28.8K	23.38 seconds
33.6K	20.12 seconds
56K	15.89 seconds
ISDN 128K	5.95 seconds
T1 1.44Mbps	1.83 seconds

4.7 SUMMARY

User satisfaction in working with Web-based information publishing depends on the quality of Web sites. The technology is available to develop, publish, search, and analyze Web sites. Corporations have been collecting a lot of experience with how Web content is accessed and used. In most cases, Webmasters and Web administrators prepare blueprints and guidelines for site development and deployment. Web sites should be continuously evaluated for optimal performance. This can be done by deploying site evaluation tools or by outsourcing this activity to a service company. Independently from this choice, periodic benchmarks are recommended to compare the Web site quality with the industry average and with best practices.

5

Log File Analysis

CONTENTS

5.1   INTRODUCTION

What is Web site activity analysis and what is its purpose? Usually Web site activity reporting involves the analysis of:

- Basic traffic statistics (hits, page views, visits)
- Navigation patterns (referrers, next click, entrance and exit pages)
- Content requested (top pages, directories, images, downloaded files)
- Visitor information (domains, browsers, platforms)
- Fulfillment of the Web site's objective (purchases, downloads, subscriptions)

Clearly, this last item is the reason Web site activity analysis has become a critical priority for organizations investing massive amounts of time and money in their Web presence. How well the Web site is performing relative to its objective is what justifies continued investment. The easiest way to quantify the return on investment (ROI) is with meaningful Web activity reports.

Reporting is also essential for making decisions about content. Web site activity reports, by providing statistics about the most popular pages or files, give an organization quantifiable measurements as to what type of content appeals to its audience. Without reliable, comprehensive reports, a Web site's content is designed based on an educated guess by the design team or editorial staff.

Similarly, Web site activity analysis reports also tell an organization about visitors to the site. Where are they coming from, how do they get to the Web site, and what type of browser or platform are they using? When a corporation decides to deploy a Web site, it usually has an idea about who its audience will be. Does the actual audience resemble the predicted one? How does it change over time? What type of content improves visitor retention or session depth? All of these questions and many more will be answered in this chapter.

5.2   USAGE ANALYSIS

Web server monitors and management tools concentrate on how the Web server is used and how performance goals can be met. In addition to these tools, other tools are required that are able to continue the analysis using log files filled by special features of the server operating system. This chapter is devoted to log file analyzer tools that are able to give the necessary data for in-depth usage analysis.

Usage analysis is a means of understanding what is happening on an Internet or intranet server such as a Web server. Usage analysis tools piece together data fragments to create a coherent picture of server activity.

Usage analysis can answer the following questions:

- How many individual users visited the site on a particular day?
- What day of the week is the site busiest?
- How many visitors are from a certain country?
- How long do visitors remain on the site?
- How many errors do visitors encounter?
- Where do visitors enter and leave the site?

- How long did it take most visitors to view the home page?
- Which links on other sites send the most visitors to this site?
- Which search engines send the most visitors to this site?

Reports can span any length of time, making it possible to see trends. They can also display any degree of granularity, allowing users to see both broad-ranging reports and detailed reports. Usage analysis is most frequently thought of in terms of Web servers. The reports created by usage analysis tools can be used throughout organizations to help people make informed decisions. For example:

- Web developers use these tools to gauge the effects of site design changes. Using this information they can make further refinements to the design of the site to maximize its effectiveness.
- Marketers use these tools to analyze the effectiveness of marketing programs and online ads.
- Site administrators can spot Web pages that are causing errors, determine future server hardware needs, and track FTP and Proxy server activity.
- Salespersons can gather information about prospects, including their geographic location, how many pages they viewed, and how they found the site in the first place.
- Executives use the intelligence gathered with log file analyzers as a resource when making a broad range of decisions.

Each time a visitor accesses a resource on a Web server — whether it is an image, an HTML file, or a script — the activity is usually recorded as a line in a text file associated with the Web server. This text file is known as the Web server log file. A single line of a typical Web server log file can be interpreted as follows:

Record of the server log file entry:
    foo.bar.com—(31/Oct/1998:23:31:44+ 500) "GET home.html HTTP/1.0"
        200 1031 http://www.yahoo.com/
    "Mozilla/3.0 (Win32;U)"
Interpretation by elements:

foo.bar.com	Host name of the visitor's computer
31/Oct/1998:23:31:44	Date and time
GET	Method used to request the resource
home.html	Name of the requested resource
HTTP/1.0	Protocol used to request the resource
200	Status code "200" means that the request was successful
1031	Number of bytes transferred to satisfy the request
http://www.logfile.ana.html	Web page that referred the visitor to this page
Mozilla/3.0	Visitor's Web browser and version
Win32	Visitor's operating system

Most Web servers write out log files in the combined log format. This differs from the older common log format in that it contains browser and referral information. Referral information is important to determine what sites are sending the most traffic to the target address and what sites might have out-of-date links pointing to specific user sites. Referral information is also critical for gauging the effectiveness of online ads. Other information that can be present in a log file includes:

Cookie	A persistent identification code assigned to a user that allows the user to be tracked across several visits
Session identifier	Tracks each visitor for the length of the visit only
Amount of time it took to fulfill the request	Enables server performance reporting

Basically, there are two types of usage analyzer tools: software-based and on-the-wire collectors. On the high end of usage analysis tools are packet sniffers, which offer on-the-wire reporting by installing an agent against the kernel of the operating system of the Web server. They run as root in the kernel of the operating system on the Web server. Furthermore, they require that a network run in promiscuous mode in order to expose network traffic to the agent. Usually, there are very few reports packet sniffers can create and log file analyzers cannot. Log file analyzers can create reports on the usage of secure/encrypted communications, which packet sniffers cannot. Packet sniffers are more expensive, offer fewer reports, and offer just a few report distribution capabilities.

5.3   ISSUES OF LOG FILE ANALYSIS

When selecting products, there are a number of criteria that must be carefully evaluated. These criteria are also important when Webmasters want to position log file analysis within their IT administration or when they want to deploy this functionality within their organization.

Architecture of a product answers the question of whether the product can support a distributed architecture. Distribution means that collecting, processing, reporting, and distributing data can be supported in various processors and at different locations. Figure 5.1 shows these functions with a distributed solution.

In Figure 5.1, Web servers A, B, and C can be from very different types, such as

• Netscape Navigator
• Microsoft Explorer

Of course, it is expected that many different Web server types are supported. Also, the hardware and operating system may be differentiators for products. It is assumed that the Web server hardware has a decreasing impact on log file analysis. The role of operating systems is more significant; the product should know exactly how log files are initiated and maintained. No problems are expected with leading Web server solutions, based on Unix and NT.

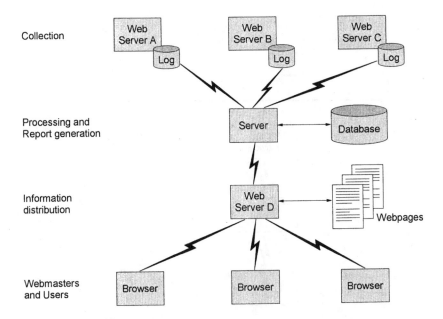

FIGURE 5.1   Generic product architecture for log file analysis.

The data capturing technique is essential with log file analysis. The first question is where the logs are located. Figure 5.1 indicates that they are located in the Web servers. However, more accurate information is required here:

- What memory area is used?
- What auxiliary storage area is used?
- What is the size of those areas?
- What types of log files are supported?

If log files are not processed in real time or near real time, it is important to know where they are stored until they are downloaded for processing. Log file analysis deals with very large data volumes, and these volumes depend on visitor traffic.

Many vendors put a lot of emphasis on how quickly a log analyzer can scan data. The speed issue is not really what it appears to be. Most products can process multiple megabytes per minute under ideal conditions. In the real world, however, initial log-read speed is often a factor of disk drives' speed not the log analyzer program, because much of the work involves reading huge amounts of data off the drive. Benchmark results are extremely good with preresolved domain names. Most vendor benchmarks don't perform reverse DNS lookups on IP addresses. Under typical conditions, it takes time for a Web server to resolve domain names. Optimal processing speed may drop in lookup and DNS conversion cases by factors of 30 to 40.

Usually, log files are downloaded for processing. It is important to know how downloads are organized and how rapidly they are executed. As indicated in

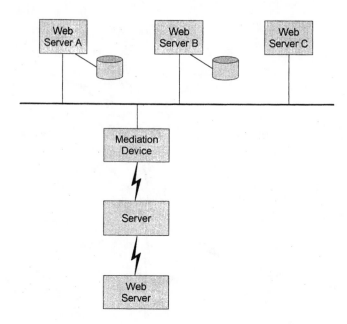

FIGURE 5.2   Use of a mediation device to reduce network overhead.

Figure 5.1, wide area networks are involved with sometimes limited bandwidth. The bandwidth is usually shared with other applications with the result of potential traffic congestion. Bandwidth-on-demand solutions are rare with log file analysis. When transmission is arranged for low traffic periods, the actuality of log file analysis results may suffer. In such cases, local storage requirements increase, and processing, report generation, and information distribution are delayed by several hours or even by days.

Two solutions may help. The first solution is to use intelligent profiling at the source of data collection. Redundant data are removed from logs during collection. Data volumes decrease and local storage requirements decrease, but processing requirements in Web servers increase considerably. The second solution is to use data compression or data compaction with the same results and impacts as with the first solution.

Overhead is a very critical issue with large data volumes. Data capturing is expected to introduce little overhead, when logs are stored away immediately. If local processing is taking place, overhead must be very carefully quantified; if resource demand is high, overall Web server performance may be affected. Data transmission overhead can be heavy, when everything is transmitted to the site where processing is taking place. WAN bandwidth is still very expensive to be dedicated just to log file analysis. If bandwidth is shared with other applications, priorities must be set higher for business applications than for transmitting raw log file data.

In the case of server farms, a local mediation device could help. Figure 5.2 shows this alternative. The mediation device is connected via LANs; bandwidth is not so critical in LANs in comparison to WANs. Processing and report generation remain at a special server that consolidates all data from mediation devices.

It is absolutely necessary that all data are captured that are necessary to conduct a detailed Web site analysis of visitors or groups of visitors:

- Who is the visitor?
- What is the purpose of the visit?
- Where did the visitor come from?
- When did the visit take place?
- What key words brought the visitor to the site?
- What search machines helped to access the site?
- How long was the visit?

Comparing different products against the same logged data, results may be different. The reason is that different tools make different assumptions during log analysis. One problem occurs with a unique visitor. Without cookies, log files don't record people or machines, but IP addresses only. If a user is browsing from behind a firewall, the program records the IP address of the proxy server. By assuming that one IP address equals one user, these programs risk underreporting visitors if more than one person accesses a site through the same proxy server.

There is also a question of how log file analyzers define a visit or session. Because HTTP is a stateless protocol, they can't tell when a user is done looking at a site, which makes it difficult to determine visitation lengths. How much inactivity should signal the end of a visit? The default setting varies from vendor to vendor.

Standardizing log definitions would improve consistency among products. BPA International (BPAI) is one organization helping to set some standards. However, although many tool vendors claim to support BMP standards, products may not return the same results without careful tuning — one more reason users need to investigate the assumptions their products make and the terminology used in the reports. Still, as long as users use the same product and settings when they analyze logs, trends can be recognized and reasonable conclusions can be made.

Data losses cannot be completely avoided. Logging functions of Web servers, storage devices, or components of the transmission may fail; in such cases, there will be gaps in the sequence of events. Back-up capabilities may be investigated, but IT budgets will not usually allow too much spending for backing up large volumes of log file data. In the worst case, certain time windows are missing in reporting and in statistics. Those gaps may be filled with extrapolated data.

Furthermore, the management capabilities are very important. One of the functions here includes automatic log cycling. In order not to lose data, multiple logs are expected to be used. When one of the logs is full, the other log seamlessly takes over. Another function is the translation of DNS (domain name service). Its speed is absolutely critical for real-time information distribution. To generate more meaningful reports, results of log file analyzers must be correlated with other data sources. These other data may be maintained in other databases. In order to correlate, ad hoc database links should be established and maintained. Management of logs of any log file analyzer can be taken over by the operating system of Web servers. The basic services are supported today; additional services may follow. In the case of server farms or of many individual Web servers, the coordination of log transfers and processing is no trivial task. An event scheduler may help in this respect.

Cookie support is important to speed up work initiated by visitors. It is a logical connection between Web sites and browsers; a persistent identification code is assigned to a user, which allows the user to be tracked across several visits.

Similar to the processing speed issue is the question of how big a log file a product can handle. Log file size is dictated more by the amount of RAM or disk space available than anything else. If a product is using a memory-based log file analysis program to crunch a gigabit-level log, it may not work well depending on the quality of the operating system in the server. Vendors prefer to use databases to store the log file data. This server two purposes: it allows users to requery the data, and it allows users to create indexes and summaries without keeping the entire log file in memory at once.

Database managers would then offer a number of built-in features to maintain log files. Visitors may be clustered according to geography, common applications, common interests on home pages, date, and time of visits. Automatic log cycling can also be supported here by the database managers. Open database connectivity (ODBC) support helps to exchange data between different databases and to correlate data from various databases. Besides log files, other data sources can be maintained in the same data warehouse. In addition to routine log file analysis with concrete targeted reports, a special analysis also may occasionally be conducted. This special analysis, called data mining, can discover traffic patterns and user/visitor behavior. Both are important to sizing systems and networking resources.

The down side of using a database is that loading the database with data can be a time-consuming procedure. Straight memory-based log file analysis tools will beat the database tools on the first pass, but these memory-based tools require a complete reread of the file to run a new query. For any decent amount of logged data, using a database is the only way to go and should help with the size issue. In order to lower disk space requirements and to reduce query time, most products dump certain portions of the log file data from the database. Despite such steps, log files are usually big. If users want to save and be able to run queries for a longer period of time, a data warehouse solution with data mining capabilities is the most promising solution.

One of the most important questions is how log file analysis performs when data volumes increase. A volume increase can be caused by offering more pages on more Web servers, more visitors, longer visits, or extensive use of page links. In any case, collection and processing capabilities must be estimated prior to deciding on procedures and products.

To reduce processing and transmission load of log files, redundant data should be filtered as near as possible to the data capturing locations. Filters can help avoid storing redundant data. They can also be very useful in the report generation process. Again, unnecessary data must not be processed for reports. Powerful filters help to streamline reporting.

Not everything can be automated with log file analysis. The user interface is still one of the most important selection criteria for products. Graphical user interfaces are likely, but simple products still work with textual interfaces. When log file analyzers are integrated with management platforms, this request is automatically met by management platforms.

Reporting is the tool to distribute the results of log file analysis. Predefined reports and report elements as well as templates help to speed up the report design and generation process. Periodic reports can be automatically generated and distributed for both single Web servers and Web server farms. In case of many Web servers, report generation must be carefully synchronized and scheduled. Flexible formatting helps to customize reports to special user needs.

Output alternatives of reports are many. The most frequently used solutions include Word, Excel, HTML, and ASCII. There are also many choices for report distribution:

- Reports may be stored on Web servers to be accessed by authorized users who are equipped with universal browsers.
- Reports can be uploaded into special servers or even pushed to selected users.
- Reports may be distributed as attachments to e-mail messages.
- Reports can be generated at remote sites; this alternative may save bandwidth when preprocessed data instead of completely formatted reports are sent to certain remote locations.

Documentation may have various forms. For immediate answers, an integrated online manual would be very helpful. Paper-based manuals are still useful for detailed answers and analysis. This role, however, will soon be taken over by Web-based documentation systems. In critical cases, a hotline can help with operational problems.

Log file analysis is actually another management application. If management platforms are used, this application can be integrated into the management platform. There are many ways to integrate; most likely a command line interface (CLI) will be deployed.

Table 5.1 summarizes the selection criteria for log file analyzers.

5.4 DRAWBACKS OF PURE LOG FILE ANALYSIS

Log file analysis can provide a good entry-level summary about the activities in and around Web servers. However, this technology has major problems, as follows:

The first major problem is traffic volume. As traffic levels quickly reached exponential growth rates, nightly log file downloads quickly became afternoon-and-evening, and then even hourly, downloads, because server disk drives filled with log file data so quickly. Compounding this problem was the fact that higher traffic sites needed to load-balance across several servers and physical machines, so that log file downloads needed to be done not only many times a day, but also across several

TABLE 5.1
Selection Criteria for Log File Analyzers

Evaluation criteria
Architecture of the product
Platform requirements
  Web server types
System requirements
  Hardware
  Operating system
Data capturing techniques
  Locations of logs
  What log files are supported
  Storing log file data
  Download of log files
  Log file import speed
  Reduction of redundancy (intelligent profiling)
Overhead
  Data capturing
  Data processing
  Data transmission
Components of Web site analysis
  Who
  Why
  Where
  When
  What (e.g., key words)
  How accessed (e.g., what search machine)
  How long accessed
Precautions against data losses
Management capabilities
  Automatic log cycling
  DNS translation speed
  Correlate information from external databases
  Embedding into OSs
  Event scheduler
Cookie support
  Cookie elements
  Visit calculation
  Automatic cookie support
Database management
  Use of a database manager
  Grouping capabilities
  Automatic log cycling
  ODBC support
  Data warehouse
  Drill-down (mining) capabilities

TABLE 5.1 (continued)
Selection Criteria for Log File Analyzers

Database management (continued)
   Quicklist database
   Query string parsing
Scalability
Flexibility
Filtering
   Reports
   Information within reports
User interface
   Textual interface
   Graphical interface
   Integration with management platforms
Reporting
   Predefined reports
   Use of templates
   Predefined report elements
   Automated server farm reporting
   Trend reporting
   Use of a report scheduler
   E-mail reports
   Report output alternatives (Word, Excel, HTML, ASCII)
   Flexible formatting
   Cross-tabs by correlating fields in the database
   Search keyword reporting
   Reporting speed examples
   Customization capabilities
Distribution of reports
   Accessing reports
   Uploading reports
   Remote reporting
Documentation and support
   Integrated online manual
   Paper manual
   Help desk
Integration with management platforms
Vendor
   Number of clients
   Founded when
   What other products are offered
   Support (e.g., on site, hotline)
   Maintenance contracts
   Financial strength
   Keeps current with servers software releases

machines each time. The quick fix to this problem was typically an automated script that would download log files on a preset schedule. However, this failed to account for unexpected spikes in traffic and also clogged internal networks with huge log files being transmitted across the network several times a day.

The second major problem is data processing speed. Even if there were an easy way to continuously transfer log file data to a consolidated area, there was still the problem of how to process the gigabytes of log files into database tables in an efficient, continuous, and robust manner. Batch processing of log file data required a considerable amount of time. In addition, the human resources demand for log file collection, processing support, and report compilation has exceeded the expectations.

The third major problem involves incomplete data. Besides log files, there are significant alternative sources of site activity data that contain more information than even the longest, most complex custom log file format can provide. A log-file-only approach cannot guarantee a complete picture for Web activities. A good example of missing data is certain network-level data that the Web server and the server's log file never get to see. For instance, a visitor requests a page that turns out to be too slow to download and decides to hit the browser STP button or BACK button or otherwise terminate the request mid-download. In this case, the network layer will log that action, but it will not notify the Web server about it. Similarly, a great deal of data that is seen by the Web servers is never written to the log file. Therefore, any measurement approach based solely on log files would occasionally miss critical information about user activity on the Web site.

The fourth major problem with the log file approach is flexibility. As sites become more sophisticated, one of the first obvious enhancements is to add dynamically generated content. Regardless of the type of content management system used, dynamic content typically results in URLs that are very difficult, if not impossible, for a human reader to decipher. Since log files are just transaction records, dump reporting systems simply pass the nonsensical URLs through to the end user report as the page that was requested, resulting in an unintelligible report with meaningless page names and URLs. The ideal solution would be to interpose some intelligent classification system between the raw activity data and end user report. In practice, however, the reality of gigabytes of raw log files often leaves an in-house analysis team with few human resources to add even more complexity to an already slow log-based process. The inflexibility of log files to handle the tracking of new technologies has been observed not only with dynamic content but also with personalization applications, applet-based multimedia technologies, and a host of other new capabilities that the log file approach was never designed to handle.

In summary, although log files were a convenient approach to measurement in the early days of the Web, they rapidly highlighted problems of:

- Labor intensity
- Slow data processing speeds and turnaround times measured in weeks
- Incomplete data; missing server- and network-level data
- Ineffective tracking of new feature enhancements such as dynamic content, personalization, and applet-based multimedia

In response to these problems, hybrid products have been developed and deployed.

5.5 TOOLS FOR LOG FILE ANALYSIS

Log file analysis for Web servers is a relatively new area. There are a number of products that address log file analysis. However, the boundary between products is very difficult to draw. Some of them focus on just analyzing log files; others include content optimization as well. Two hybrid products, SiteServer from Microsoft and ARIA from Andromedia, are analyzed in Section 5.6.

5.5.1 NET.ANALYSIS FROM NET.GENESIS

net.Analysis Pro has the scalability and analytical power required for analyzing the most advanced Web sites. It also provides tools for distributing key Web site data enterprise-wide. It contributes to the success of any Web site by providing actionable information to Webmasters, content owners, online marketing managers, and anyone else needing site information. In addition, it automates all aspects of report production and distribution, easing the burden of Web site reporting on the corporate Webmaster or IS administrator.

Key features are:

- More than 100 predefined reports for the different information needs of Web site administrators, marketers, and other members of the organization; also, different filters can be applied
- KeyValue analysis for investigating dynamic site content, search engine keywords, and advertising click-throughs
- ReportSite, a ready-built intranet site where reports can be published and accessed
- Automated publishing system which automates log file handling, report production, and distribution
- HTML reporter, which allows anyone in the organization to choose and run reports from a Web browser
- Analyzes FTP, proxa, Lotus Domino, and streaming media servers in addition to all common Web servers
- Full support for large and complex Web server configurations

net.Analysis Pro works by first reading a Web server's log file into the net.Analysis engine. It then processes the data, turning hits into visits, host names into the names of organizations with complete postal addresses, and Web page URLs into plain English document titles. In addition, net.Analysis ensures that reports will be available quickly by precalculating key data summaries. These summaries, called aggregates, are the key to net.Analysis being able to run usage analysis reports quickly. The engine stores the information in a robust commercial database called the DataStore. All administration is accomplished through the Web-based administrator console.

The net.Analysis Pro NT engine also offers the automated publishing system (APS), which allows administrators to automate common tasks, including the running of daily, weekly, and monthly reports. The APS publishes reports either to the ReportSite — a built-in and self-administering intranet site — or to any other location. Users browsing the ReportSite have the option of self-describing to reports that interest them. Reports are then pushed to them via e-mail each time they are generated.

Site administrators use the Windows or NT reporter to build and run reports. This application offers a host of report-customization features and also allows users to drill down into report results to explore underlying trends. Less sophisticated users run prebuilt reports through a Web browser using HTML reporter.

Each time visitors enter a Web site they leave a trail of data behind, even if the Web server does not tag them with a cookie. In order to string a group of clicks together into a visit, net.Analysis does the following:

- First it looks for a user identifier, which can be either a cookie or a user ID if the site requires user registration. If a user identifier is present, net.Analysis can easily group all of the accesses by the user together as a visit. If a user has not accessed anything in 30 minutes, the visit is considered ended.
- If it does not find a user identifier, net.Analysis looks for a session identifier. This expires if the visitor has not accessed anything on the site within 30 minutes. By stringing together all of the visits with the same session identifier net.Analysis can easily create a visit.
- If there is no session identifier, net.Analysis looks at the user's host name. Because two visitors using the same ISP could have the same host name, net.Analysis looks for a series of clicks from the same host name, browser version, and operating system. Once these conditions are met, net.Analysis creates a visit. As before, if the user has not accessed anything in 30 minutes, the visit is ended. In addition, the visit will end if net.Analysis detects that the visitor has left and reentered the site.

net.Analysis offers full error status reporting, and it can identify any bad links that visitors encounter on a site. If there is a bad link on a site that a visitor has not encountered, however, net.Analysis will not be able to report on it.

net.Analysis supports the following Web servers:

- Netscape Enterprise
- Microsoft IIS
- O'Reilly Web site
- Open Market
- Zeus
- Webstar

The performance of Web servers can also be analyzed. net.Analysis provides information on how long it takes to access and download information from the Web server. It reports the result using a number of prebuilt analysis reports. net.Analysis runs on Windows NT 4.0+ and Sun Solaris 2.4 or 2.5. net.Analysis can run anywhere on the network of the user. If the product resides on the Web server, reports should be scheduled at a time site usage is expected to be low.

net.Analysis does more than generate static reports. When users click on any line of a net.Analysis report they are presented with a list of available queries that are key to exploring underlying trends. This capability is much more powerful than simple, linear drill down because it lets users navigate through site data.

In summary, net.Analysis is a powerful log file analyzing tool that answers advanced questions, such as:

- How many pages do visitors typically view on a specific Web site?
- How much time do visitors spend at a Web site?
- What are the most common paths to the most important pages on a specific Web site?
- Which parts of the Web site are most popular with Fortune 500 and Fortune 2000 companies?
- What keywords bring the most traffic to the site from search engines?
- How long does it take visitors from a certain state of country to view particular home pages?
- How many new users visit the particular home page?

5.5.2 NetIntellect from WebManage

WebManage Technologies develops and markets software solutions that can deliver predictable services to high-traffic Internet and intranet sites, e-commerce sites, and ISPs. Using new technology and open industry standards, WebManage has delivered one of the first Java-based comprehensive, centralized, and open implementation of policy-driven site resource management (SRM) and class of service (COS) over the Web at the application layer.

NetIntellect is one of the essential tools for Internet and intranet management, marketing, sales, customer service, and server administration departments. NetIntellect is a typical Web server log analysis tool that generates reports that reveal critical statistical, geographical, and marketing trends on Web sites.

The main features of the product can be summarized as follows:

Support of decision making:
- How do Web site changes affect the Web site traffic?
- Is the intranet advertising campaign increasing the Web site traffic?
- What are international, national, and local users looking for when they visit the Web sites?
- When is the peak usage time for particular Web servers?
- What are the postal and e-mail addresses of the hosts visiting the Web sites?
- Which pages appeal the most to visitors?
- How much time do visitors spend on each page?

- What is the activity by market segment?
- Which are the most popular products or services on particular Web sites?
- Which hour of the day or day of the week is the best to advertise?
- What are the hourly, daily, and weekly trends on the Web sites?
- Where do users come from?
- Which are the top referring sites and URLs?
- How many bytes of data are being retrieved from particular Web sites?
- What are the frequently used visitor platforms and browsers?

Analysis of logs:

- It has an intuitive user interface.
- It allows for automatic identification of log file formats.
- Log files can be accessed remotely with the built-in FTP; the reports can be automatically uploaded to an FTP/Web server and/or saved to a specific directory.
- Log file manager allows users to quickly obtain summary data, manage log files efficiently, and add or delete logs. The "FIND LOGS" feature lets users easily locate and add log files to the log file manager.
- Multiple log file selection allows users to select and process multiple log files in one run and display results in a single report.
- The "APPEND" feature allows users to append/combine different log files to generate one report and/or gather historical data.
- It optimizes disk space by creating databases that can be less than 2% of the size of the original log files. This allows users to store logs for an entire year in one database file.

Scheduling options:

- A built-in scheduler allows users to automatically generate multiple reports at any time in HTML and MS Word formats and e-mail them directly to the customers, managers, and administrators.
- The scheduler options console allows users to define options for an unlimited number of entries; each can create a report on a different or the same log file with a unique set of reporting options.

Reporting:

- Users can select from more than 100 predefined colorful reports that reveal statistical, geographical, and marketing trends.
- Users create and save any number of custom reports and apply more than 50 pre- and postprocessing filters.
- Besides the default display, reports can be generated and viewed in HTML or MS Word format; users can even generate MS Access database files.
- Users can create individual reports for multiple domains and virtual servers, even if they have just a few home pages.
- The unique drill-down feature lets users explore the entire log file and concentrate on important data.
- Allows mapping of the whole intranet by generating reports showing activities by organizational units.

In particular, the identification of Web site visitors is valuable. NetIntellect can translate IP addresses to domain names and complete company names. The built-in organizational database provides postal and e-mail addresses as well as contact names of the hosts visiting the particular Web site. This database also includes an international listing.

SiteMARC allows network managers and site administrators to build highly scalable, available, reliable, and responsive Internet and intranet sites. Using SiteMARC e-commerce, corporate presence and Web-based customer support sites can deliver services commensurate with users' subscribed service level and class of service. Site administrators can group users into groups and provide access privileges accordingly. SiteMARC provides the following services in one package:

- Rules-based intelligent traffic distribution (load balancing)
- Privileges-based site access, based on users' service level and class agreements
- Realm-based access control to allow or deny access to site content
- Real-time monitoring of servers and processes
- Site content management and redirection
- Browser-based site integrity analysis and site mapping
- Browser-based real-time site activity reporting
- Browser-based centralized site administration

WebManage products protect customers' legacy and new investments in site hardware and software, ensure multivendor interoperability through platform-independent implementation, and support incremental deployment.

5.5.3 WebTrends Products

WebTrends is a C/C++ application that can run on any Windows-based (95, 98, NT 4.0) system in the network. No extra servers or software are needed. Built-in remote reporting capabilities allow the log analyzer to function as a reporting server with a Web browser client, but neither are required to run the software.

The product suite consists of four members:

- Web traffic and log analysis
- Proxy analysis
- Web link analysis and quality control
- Monitoring and alerting

WebTrends is a 32-bit application developed and compiled in the C/C++ programming language for speed, portability, and industry standards compatibility. It does not require the importing phase other log analyzers require; therefore, it is faster, easier to use, requires a reasonable amount of disk space, and does not have the usual memory limitation of other applications. WebTrend stores analysis results in a database so subsequent analysis of the same timeframe requires a shorter run time. The WebTrend Log Analyzer can process 40 million hits a day and log files larger than 10 gigabytes.

Log files may reside on any server in the system that the WebTrends Log Analyzer can access through HTTP, FTP, local file access, or UNC (Universal Naming Convention) mapped drives.

More than 30 log file formats, including Microsoft FTP, Proxy, Firewall, and Web server log files are supported by the product. The WebTrends Log Analyzer supports the above plus the following formats: Apache Log Format, Apache Extended Log Format, Best Internet Log Format, CERN Log Format, Cold Fusion Log Format, Emwac Log Format, HP Log Format, IBM Internet Connection Secure Server Log Format, Lotus Domino Log Format, Market Focus Log Format, MCI Log Format, Microsoft IIS Extended Log File Format (MS IIS), Microsoft IIS International Date Format (MS IIS), Microsoft Internet Information Server 3.0 (MS IIS), Microsoft Internet Information Server 4.0, Microsoft ODBC Log Format, NCSA Combined (Extended) Log File, NCSA Common Log File, NCSA Common Multi Home, Netscape Enterprise Server, Netscape FastTrack Log Format, Netscape NSAPI Extended Format, Purveyor Extended Format, Purveyor Multi-Home Format, Open Market Log File, Oracle Log Format, Real Audio Log File, Spry Log Format, WebSite Combined Multi-Home Format, WebSite Common Multi-Home Format. The suite products also support NCSA Proxy, Netscape Proxy Log Format, Novell Border Manager, and Microsoft Proxy Log Format.

Most features are present with this product, including users listed by region and summary of activity by day, day of the week, and hour. WebTrends also gives the user feedback on bandwidth utilization and server statistics, such as number of cached and failed links. Any log file formats can be handled. In particular, Microsoft IIS can be analyzed in depth. However, special DLL (Dynamic Link Library) files must be installed on the servers the user wants to analyze. Log files must be selected prior to processing. Locating and selecting the log files to process is not as easy as would be required by users. There is no scanning function with this product. Grouping log files together for single processing and reporting is difficult. These inconveniences are outweighed by the processing capabilities. Report attributes can be selected before or after processing. Before processing, the user can determine whether or not to convert numeric IP addresses to domain names and if it should store data in the FastTrend database. FastTrend allows users to store log information in a database instead of rereading the log file each time the user wants to create a report.

WebTrends Log Analyzer can break down users by IP, machine name, authenticated user name, country, state, city, organization, domain type, intranet departments, etc. Each piece of information is presented in its own table/graph (top users, for instance) and is also mixed into other tables/graphs.

WebTrends products can identify visitors' organizations, the links that led them to the site, how long they stayed, what pages were visited and in what sequence, and what information was downloaded, and with filters, how the quality of users from one referring link or ad compares to others. In particular, questions like what, how long, and how accessed are interesting. Log Analyzer offers the following answers:

What (e.g., keywords) and how (e.g., what search machine) accessed: Web-
Trends products report on referring URL, referring site or engine, top
keywords, and top keywords by engine. It also has ad tracking abilities,
which means it can trace views and clicks of advertising on your site.

How long accessed: Tables listing specific Web pages include an "average
time viewed" column.

Also included are top entry pages, top exit pages, single access pages (users view
this page, then leave), top paths through the site, and many more. The best bet is to
look at a report and refer to the features list at: http://www.webtrends.com/prod-
ucts/log/info.htm.

IT and network managers are interested in getting answers for the following
questions:

How can you determine the return on your investment?
How can you measure the traffic on your Web server?
Can you measure both quality and quantity of visitors to your Web site?
How many users visit your site daily? Is that number growing?
What paths do visitors take when they browse your Web site?
From what countries do users connect? What cities? What states?
From what departments do users connect to your intranet servers?
Which is the most active day of the week? The most active hour?
What kind of information is accessed on your server?
Which pages are the most popular?
How many pages are accessed in each directory?
How many users access each directory?
What browsers are used to access your Web server?
What operating systems are used in Web servers?
Which forms are submitted most often?
How are intranet Web servers used?

WebTrends Log Analyzer offers answers for all of these questions by generating
various reports.

The user can create reports in HTML, MS Word or MS Excel, text, and comma-
delimited formats. The user can also specify what language (English, Spanish,
German, or French) in which to create the report. Furthermore, the user can create
custom reports and save them for use again later. WebTrends also comes with
support for intranet reporting, and the user can easily add intranet sites to the
database.

Each report can be made up of from one to all sections, and most sections contain
at least one table and graph. Reporting examples are:

- General statistics
- Top visited HTML pages
- Top file downloads
- Top executed scripts

- Top submitted forms
- Most active organizations
- Organization type
- Top countries
- Top states
- Top cities
- Activity by day of week
- Activity by hour of day
- Web server/technical information
- Browser/client errors
- Server errors
- Bandwidth and directory activity
- Types of files downloaded
- Proxy server information
- Most-used browsers
- Most-used Netscape browsers by version
- Most-used Microsoft browsers by version
- Referring (advertising/listing) sites
- Most-used platforms and operating systems
- Referring (advertising/listing) URL pages
- Top paths
- Ad clicks
- Ad views
- Top entry pages
- Top exit pages
- Bottom pages
- Visiting spiders, crawlers, and robots
- Top pages of single-hit user sessions
- FTP uploads (for IIS servers)
- Bandwidth utilization
- Authenticated users
- Search variables of popular search engines

For Webmasters, report processing schedules are supported as well. To build a regular update of several report types, users can define as many different schedules as needed and have them run at user-defined time intervals. To create a schedule, users must first define filters to the selected log files. Here, there is no difference between manual and automated processing.

Table 5.2 shows an example for general Web server statistics. In this reported case, the tool could not differentiate between the origins of user sessions. Figure 5.3 displays Web activities by the day, indicating very heavy Monday traffic. Figure 5.4 shows more details by breaking down the daily traffic by the hour, indicating a clear peak around 9:00 a.m.

This product has the power to let users manipulate their log files to see exactly the data they need to see. WebTrends can work with management platforms; the first implementation example is with TNG Unicenter from Computer Associates.

TABLE 5.2
General Web Server Statistics by WebTrends

Date and time this report was generated	Friday, December 4, 1998; 07:46:17 a.m.
Time frame	11/01/98 01:03:38 – 11/30/98 22:05:41
Number of hits for home page	1031
Total number of successful hits	5729
Total number of user sessions	1119
User sessions from the United States	0% (not broken down)
User sessions from outside the United States	0% (not broken down)
Origin unknown user sessions	100%
Average hits per day	190
Average user sessions per day	37
Average user session length	67:11:49

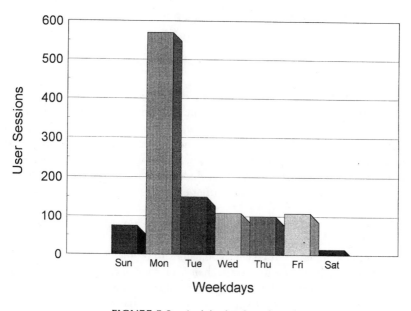

FIGURE 5.3   Activity by day of week.

5.5.4   INSIGHT FROM ACCRUE SOFTWARE

Insight is a comprehensive Web site analysis solution that helps users optimize the effectiveness of electronic commerce. Capable of managing rapidly growing and enterprise-class Web sites, it allows managers to profile their online customers, understand how visitors move through their site, determine what content is being viewed, and strengthen online marketing campaigns. Web site managers and marketers use these new sources of information to lower user acquisition costs, increase visitor retention, create effective marketing programs, and lower infrastructure maintenance costs.

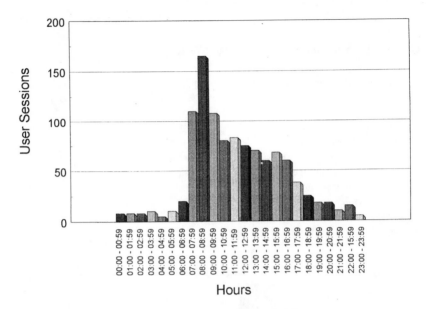

FIGURE 5.4   Activity by hour of day.

A database and intuitive user interface create an easy-to-use system for organizations to generate, evaluate, and distribute meaningful data about Web site visitors. Reports can be structured to meet the specific needs of the businesses, departments, or projects, providing a detailed view of the activity at any Web site. In addition to existing report options that permit in-depth analysis of user activity, content, delivery, host, and domain information, Insight includes new reports that provide valuable data about the visitor experience. Reporting examples are:

- Referrer report — It compiles detailed information about Web sites generating traffic to a particular site.
- User type report — It enables companies to profile the unique characteristics of visitors who transact at their Web sites, including which visitors are registered or unregistered, the number of repeat visitors, and the number of new users arriving in a given time span.
- Entry/exit report — It highlights individual URLs, providing information about where users enter and exit particular sites.

All these reports can support user acquisition campaigns, retention studies, and conversion programs. Reports can be exported into a spreadsheet. Printing and electronic distribution of reports can be accomplished easily as an HTML table as well.

Technical approach: Insight is a unique, network-based Web solution designed to provide a complete picture of the online visitor experience with Web sites. It allows companies with multiserver, multisite configurations to track user activity across the various servers as a single session. Through an application programming interface (API), Insight introduces support for dynamic content. The ability to handle encrypted transactions is also supported. All Web site traffic data, regardless of the complexity of the Web site configuration, is tracked and stored in a data warehouse. Through a Java-enabled browser, authorized users can access the information as well as generate and analyze critical business and marketing data. Live reports accommodate time-sensitive analysis of event-based marketing programs. Automation of routine tasks provides site administrators with increased system flexibility and performance. Capable of managing large amounts of data, Insight's data warehouse can store gigabytes of data. Through the use of filters, groups, and access control mechanisms, it can consolidate information into a predefined set of reports and deliver them to authorized users.

Insight provides marketing and Web professionals with the information they need to improve the effectiveness of their online presence. An important resource for any company that considers the Web as a strategic asset to their business, Insight provides solutions to effectively leverage Web site presence. Accurate and targeted analysis of Web site activity ensures that corporations maximize the return on their online investment and effectively achieve their business objectives.

Key features are:

- Advanced network-based data capture — Network-based on-the-wire collectors record extensive information, including delivery and performance characteristics (e.g., resets, requests, and download time).
- Enterprise-class performance and scalability — Insight is designed for complex, distributed environments composed of single or multiple Web servers. With a data warehouse for performance and scalability, it can process more than 10 million hits a day.
- Highly accurate data — The reports are not only the pages a visitor requests, but whether or not that information is actually delivered.
- Centralized data consolidation – Data from multiple or mirrored sites, running on any Web server platform, is consolidated into one central repository. This unifies the support of several sites, simplifying the collection process and freeing users from the time-consuming constraints of downloading and analyzing multiple log files.
- Enterprise-wide access — The distributed architecture allows report viewing with a standard Java-enabled browser from any local or remote location. Any authorized users can access the information they need quickly.
- Security — The administrator defines the security levels, allowing only authorized users to access all, or a subset, of the sites and site areas available.
- Quick report generation — Automated report scheduling features enable users to generate reports during off hours.
- Powerful report filtering — Specific IP addresses, hosts, domains, and URLs can be grouped and analyzed as independent entities. This allows users to generate customized reports tailored to their needs and interests.

- Exporting capabilities — Reports can be exported into several formats including HTML, Excel, and comma-separated values (CSVs). Users can then easily customize and distribute reports.
- Live reporting — Insight tracks data live, providing a view to the site activity as it occurs. Problems can be resolved immediately, ensuring that the visitor's experience is trouble-free.
- Targeted business reporting — Advertising funds can be effectively appropriated by determining, through business reports, which sites are supplying the highest percentage of paying customers.

Insight provides automation services through the batching of reports. This capability allows the marketing professional, site administrator, or technical expert to predefine a set of reports and have the reports generated at regular intervals. Responsiveness is offered by caching of reports. Once a report is generated, the data making up the report is stored so that any subsequent request for that report is immediately displayed. Caching improves performance for reports that are accessed frequently by eliminating the need to regenerate reports.

5.5.5 HIT LIST FROM MARKETWARE

Hit List Pro — This offers speed to analyze even the most complex Web sites. In addition to its speed, the product is complemented by 280+ report elements (calculations, graphs, tables, and text) in 28+ ready-to-run reports. It automatically identifies virtual domains, reports advertising effectiveness, and details the keywords used in search engines to find particular sites. All reports and report elements can be edited and copied and new ones created. Most important, Pro allows the flexibility of report and database activity scheduling, as well as three types of security-enhanced remote report types.

Hit List Enterprise — This is a high-end log file analysis solution for enterprise-level Internet, intranet, and proxy operations. It features a database technology called QuickList X-Treme, which processes, stores, and reports log file data at 650+ MB/h. It allows users to store to multiple Microsoft SQL server and access databases for flexibility. It also features DataLink, which revolutionizes analysis by enabling users to link the e-mail address, real name, phone number, and other marketing information into Hit List reports. DataLink connects to any external ODBC data source and uses cookies, IP, computer name, or user name as a key field. It combines datalinking with any of 310+ calculations and tables to provide the most comprehensive analysis.

Hit List Live — This is the high-end Web mining solution for enterprise-wide, mission-critical Web and proxy operations. Live features a unique on-the-wire technology that allows monitoring multiple NT, Unix, and Mac servers in real time. Live data collectors store to local databases as well as a global data warehouse via LAN, WAN, or the Internet. Live's QuickList X-treme database technology allows the flexibility of reading log files as a fail-safe in the event of a network or hardware failure. It also features DataLink. DataLink enhances any of 310+ calculation elements to provide the most comprehensive Web mining available.

5.5.6 NetTracker Product Family from Sane Solutions

NetTracker is an Internet usage tracking program that allows marketing professionals, Webmasters, and ISPs to get the essential information they need to make informed decisions regarding their Web sites.

The following questions may be answered by using this product:

- Who is visiting the Web site?
- Where are visitors coming from?
- When are they coming to particular Web sites?
- How long are they staying on particular Web sites?
- What pages are they viewing?
- What Web browsers are they using?
- How are people finding particular Web sites?
- What keywords are people using to find particular Web sites in the search engines?

Anyone with a Web browser can use NetTracker. Thus, reports are accessible to nontechnical personnel, too. Just in case, a context-sensitive help function is always available and can be accessed via the Web browser interface. NetTracker is installed on one of the user's Web servers and accessed via the Web browser. The product can be run by many users at the same time, from any location.

The features and benefits of NetTracker include:

- Resides on the server — Multiple users can use NetTracker simultaneously. There is no need to download large log files. It needs to be set up and configured once. Users need to purchase only one copy.
- Availability of reports — It provides users with a quick overview of data. It organizes and presents data in a clear and intuitive manner. All customized reports can be generated and saved.
- Access via Web browsers — NetTracker can be accessed anywhere, by anyone with a universal Web browser.
- Drill-down capabilities — Users can get as detailed information as they desire presented in dynamic forms.
- Use of advanced sorting functions — Users can select only the information they need by entering information in filter tools. They can also dynamically switch order ranking from views to visits.
- Data export capabilities — Data can be imported into popular software products such as MS Excel and MS Access. Data can be analyzed with other sales and marketing data.
- Context-sensitive help — Users get help information they need during operations; manuals are used very infrequently.
- Ease of use — It requires very little training and knowledge. Its setup is simple and not time consuming.

NetTracker provides a large number of reports. A few examples will be referenced here:

Date summary — This report presents information about the total number of visits and pages viewed for each month, week, and day. By clicking on the highlighted month, users can get daily and weekly breakdowns of their visits and page views. By clicking on the highlighted number of views and visits, users can get detailed information about each view and visit to their Web site during the specified day or week.

Entry summary — This report presents information about the top entry pages viewed by visitors of a Web site. By clicking on the highlighted pages, users can view each specific page. Users can also generate reports that include or exclude specific entry pages by entering the page's file name into the text box in the filter tool located at the bottom of the page. By clicking on the highlighted number of views, users can get detailed information about each page view.

Error summary — This report presents information about the top requested files or pages that visitors attempted to view, but received an error message instead. Usually, these errors occur due to broken links and unauthorized attempts to access restricted pages. By clicking on the highlighted number of errors, users can get detailed information about each error.

Exit summary — This report presents information about the top exit pages viewed by visitors of a Web site. By clicking on the highlighted pages, users can view each specific page. Users can also generate reports that include or exclude specific exit pages by entering the page's file name into the text box in the filter tool located at the bottom of the page. By clicking on the highlighted number of views, users can get detailed information about each page view.

Keyword summary — This presents the terms and words that visitors typed in at various search engines when searching particular Web sites. Users can click on the highlighted keywords and get more information about each referrer. Users can then click on the highlighted referring pages and view the last page that each visitor viewed before coming to a particular Web site. By clicking on highlighted visits, users can get detailed information about each visit to a particular Web site by someone who used the specified keywords to locate the Web site. Users can also generate reports that include or exclude specific keywords by entering the keywords into the text box in the filter tool.

Page summary — This report presents information about the top pages viewed by visitors of a Web site. By clicking on the highlighted pages, users can view each specific page. Users can also generate reports that include or exclude specific pages by entering the page's file name into the text box in the filter tool. By clicking on the highlighted number of views, users can get detailed information about each page view.

Path summary — This summary presents information about the top paths taken by visitors within a Web site. By clicking on the highlighted pages within each path, users can view each specific page. Users can also get detailed information about each visit that followed the specified path traversal by clicking on the highlighted traversal.

User summary — This report presents information about the top users of a Web site. NetTracker users can choose to have visitors ranked by the number of views or visits to their Web site. They can also generate reports that include or exclude specific visitors by entering the user name into the text box in the filter tool. By clicking on the highlighted names, users can get more information about each specific visitor. By clicking on the highlighted number or views and visits, users can get additional information about each visit and view.

NetTracker is packaged into three different products:

- NetTracker
- NetTracker Professional (Pro)
- NetTracker Enterprise

These three packaging alternatives offer flexibility for users because of their different size and complexity.

5.6   HYBRID PRODUCTS

Two more products will be introduced here. They offer more functionality than log file analysis. They not only monitor Web activity and provide historical reports, but also analyze Web content.

5.6.1   ARIA FROM ANDROMEDIA

This product line is a distributed enterprise system that captures, records, and reports on Web site activity in real time. ARIA is both an end user application for determining Web site effectiveness and a platform for delivering real-time information about users and content that other vertical applications can manipulate to drive electronic commerce.

Enterprises with high-scale complexity Web infrastructures all face the same difficulty: how to analyze the masses of activity data available to them to make sense of the activity on their Internet and intranet Web sites. Until ARIA, all in-house and third-party products took fundamentally the same approach to data analysis. Whatever the method of data capture — server logs, server plug-in, or network-level packet sniffing — they all process and store the data in a resource-intensive fashion.

The traditional approach, which may be acceptable for simple, low-traffic sites, is not efficient, and not practical for high-traffic sites and the highly complex Web infrastructures they use. For each of the thousands of page views on a typical high-end Web site, multiple log entries (sometimes ten or more) are created. The data explodes again, as each of those logs is transferred to the tracking system, where each is, in turn, parsed into several components. Those components are stored and maintained in a large relational database or data warehouse, most likely with the help of a database administrator. Once this work is done, standard queries can be run against the data, assuming the volume of data does not overwhelm the system and its users.

There are two key disadvantages to this approach for complex environments:

- It is costly to retrieve the masses of data created, aggregate it, store it and analyze it. Large installations require data mining and incur considerable database administration costs.
- Log analysis and reporting systems are unacceptably slow and CPU-intensive for large sites. The unavoidable delays in processing the data mean that the data from these systems are much less timely, and thus less valuable, particularly in the context of the rapidly changing Internet.

Similar problems exist with "live" high-end products. Although the data collection method may be different for log-based systems, the data itself and the way it is analyzed is fundamentally the same. Usually, these systems are deployed at high-end Web sites; the hardware resources are pushed to their limits. The difficulty of maintaining the products and keeping them functional as the amount of data increases exponentially is overwhelming at many enterprises. Although data capturing may be in real time, the time required to analyze and query the data grows exponentially, meaning that the data is often not usable for several hours or even days after it is collected.

5.6.1.1 Intelligent Profiling

ARIA takes a different approach to the data volume problem by using the technique of intelligent profiling. In the log analysis systems discussed above, much of the data stored is redundant. In ARIA's object database, user profiles and Web content profiles are immediately updated each time any activity occurs on the site. As a result, the information is immediately accessible for the reporter, or for any other time-sensitive application. Databases for intelligent profiling are considerably more space-efficient than for log analysis systems. On the largest Web sites, with millions of page views per day, the ARIA database can be as small as one-hundredth the size of the relational database required for analyzing log messages. Smaller databases make real-time data retrieval a reality, and database administrators' costs fall to a minimum.

5.6.1.2 Functional Modules

ARIA integrates with the most complex Web infrastructures and will scale as Web site traffic grows exponentially. Furthermore, dynamic Web site features are added. ARIA's modular platform plugs into third-party applications including ad and content management servers, and it will gracefully integrate with future technologies. ARIA's entire cycle is fully automated — from the data collection process to the time the reports are distributed in real time. Network and server monitors collect a superset of data at the server and network levels; the recorder cleans and processes that data into meaningful information; the reporter compiles and distributes reports in real

Monitors

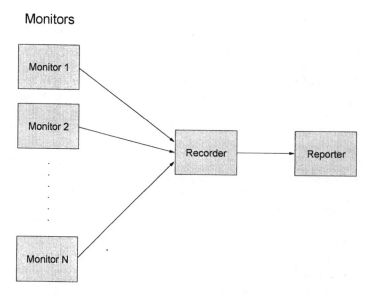

FIGURE 5.5 Architecture of ARIA from Andromedia.

time; and the exporter allows users to export data in flat file format, readable by any relational database or data mining tool.

This product can uniquely handle the problems associated with high-traffic, enterprise Web sites. There are three principal modules: the monitors, the recorder, and the reporter (Figure 5.5). Together, they represent an open, flexible, and high-performance Web activity information system.

5.6.1.2.1 ARIA monitors

The choice of which monitor type to use depends on the specific details of the site's network architecture. ARIA offers two different types of monitors, allowing it to handle high-volume and complex Web sites in a fully automated manner:

ARIA *server monitor*: This is a flexible shared library that runs dynamically on a Web server and monitors all the traffic between Web clients and the Web server. It streams to the recorder the activity information essential for understanding how a Web site is being used. Because the server monitor plugs directly into the server API, it is fully SSL compatible, meaning the user can track encrypted communications between the user and the server. It is a very small application written for the Web server API. It captures all HTTP traffic received or sent by a Web server, then sends it to the recorder for processing. Since it is a Web server plug-in, the server monitor runs whenever the server is up and running. It sends cookies to visitors' browsers, which the browsers send back to the server on subsequent hits. ARIA uses these cookies to determine when a series of hits comes from the same visitor and when a visitor returns to the site after an absence of more than a specified time period and thus begins a new visit.

ARIA network monitor: This is a network-level packet sniffer that captures all Web site activity at the HTTP level. As a result, in addition to the information available from the server monitor, the network monitor can collect network-level information, such as the number of times visitors interrupted downloads of certain pages, by hitting the stop button or making another server request before the content was fully downloaded. It includes the network monitor component itself and the ARIA relay. The network monitor scans all network traffic to and from the machine where the Web server is running. It extracts the server's HTTP traffic from the network traffic and can gather information, such as the title, from HTML pages. The ARIA relay compresses the data gathered by the network monitor and sends it to the recorder. The network monitor must be installed on the same machine as the server it monitors.

The ARIA relay can send data to the recorder using either push or pull. For installations, where the monitored server is outside a firewall and the recorder is inside, the pull configuration has the advantage of not requiring that the ARIA relay machine be allowed to send data through the firewall.

ARIA's network monitor can detect cookies set by Web servers and use them or other forms of user ID to track user paths and identify repeat visitors. It does not set cookies. If no user ID is available for a visitor, the network monitor creates one from the visitor's IP address and browser information. It can also detect POST data submitted from forms by visitors, whereas the ARIA server monitor cannot.

5.6.1.2.2   ARIA recorder

The ARIA recorder accepts the data streams of Web site activity provided by the ARIA monitors. The recorder can run either locally on the same machine as the Web server or on a different machine. Designed for high performance and scalability, it manages efficient processing during peak surges of activity. A single recorder engine can accept data streams from multiple Web servers, each running a monitor. The recorder parses the incoming data stream and creates or updates the persistent user profiles and content profiles, as well as useful aggregated statistical information, in the high-performance database embedded in the ARIA system. The highly efficient object database format offers instantaneous object updating and retrieval for the highest performance analysis and reporting available. This real-time aspect is also what allows users to drive custom content applications based on user behavior.

The recorder accepts data from both types of monitor, buffers the data if necessary, processes it, and uses the resulting information to update the objects in the ARIA database. A single recorder can accept data from multiple monitors and create a separate database to track activity from each monitored server.

ARIA maintains information in the database at decreasing granularity as the information ages. For example, data for hourly time intervals is maintained for only one week by default, but data for weekly intervals is maintained indefinitely. Data objects can be automatically removed when they reach a certain age. Maintaining less granular data as time goes by keeps the size of the database relatively constant. The object-oriented database used by the ARIA recorder provides excellent scalability. Rather than recording each page as a series of hits, ARIA records activity by each object on a Web site, so whenever a server requests a page, the hit counter is incremented only by one. The results are highly efficient data collection, storage, and processing.

5.6.1.2.3 ARIA reporter

The ARIA reporter automatically generates activity reports for the Web sites. The reports can be password protected and are viewed in an HTML browser from anywhere on the Internet. The main features are as follows:

- Intuitive graphical reporting interface and administrative interfaces
- Parameter-driven reporting: user selects which database, report type, and time frame to use for reporting
- Automated top-level summary report generation, at user-configured intervals
- Extensive drill-down capabilities of top-level reports
- Customer-configurable groupings of users and Web content for improved analysis

The ARIA reporter requires a Web server running on the same machine to serve its reports. The following types of reports are offered:

- Traffic reports
- Navigation reports
- Visitor reports
- Content reports
- Drill-down reports

ARIA handles complexity on two fronts:

- ARIAs modular structure allows for complexity at the infrastructure level, including
  Diverse server software and hardware platforms
  Geographically dispersed servers
  Mirrors, proxies, and DNS load balancing
  Mixed SSL/non-SSL environments
  Firewalls
  Bandwidth-sensitive networks and performance-sensitive servers
  ISP-hosted environments
- ARIA monitors and reports in the face of complex Web site features, including
  Dynamic content objects
  User profile objects
  Streaming multimedia objects
  User interaction with client-side applets
  User behavior in chat rooms and virtual reality environments

Figures 5.6–5.10 show selected examples of reports.

References from http://metrix.sybase.com

Site name: www.andromedia.com

Report period: From: Tue Mar 23 1999 12:00 AM
                  To: Tue Mar 23 1999 3:59 PM US/Pacific

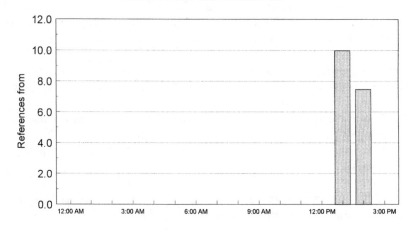

FIGURE 5.6   Visits over time by ARIA.

Visits by Content Category

Site name: www.andromedia.com

Report period: From: Tue Mar 23 1999 12:00 AM
                  To: Tue Mar 23 1999 3:59 PM US/Pacific

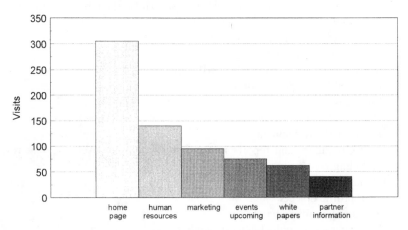

FIGURE 5.7   Visits by content category.

Visits by Visitor Category

Site name: www.andromedia.com

Report period: From: Tue Mar 23 1999 12:00 AM
                To: Tue Mar 23 1999 3:59 PM US/Pacific

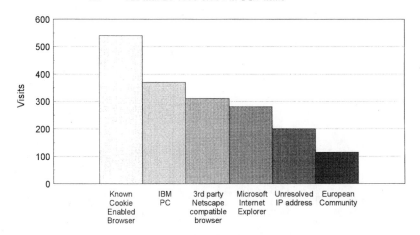

FIGURE 5.8 Hits by visitor category.

Top External References

Site name: www.andromedia.com

Report period: From: Tue Mar 23 1999 12:00 AM
                To: Tue Mar 23 1999 3:59 PM US/Pacific

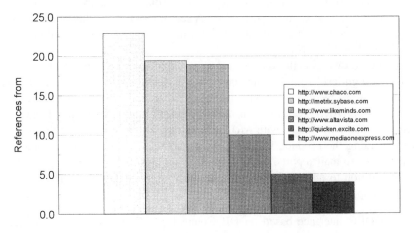

FIGURE 5.9 Most popular entrance pages.

Page Views

Site name: www.andromedia.com

Report period: From: Tue Mar 23 1999 12:00 AM
                To: Tue Mar 23 1999 3:00 PM US/Pacific

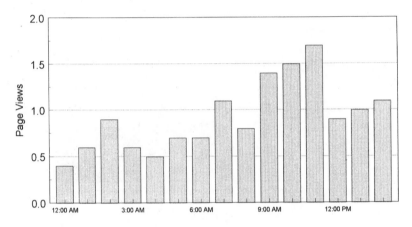

FIGURE 5.10   Page views.

5.6.1.3   Software Configuration and APIs

ARIA represents a turnkey software package. Simple instructions allow customers to download the software from the Web site of Andromedia and install it on their hardware. The software can be installed in a single-host configuration, with all ARIA components residing on the Web server machine, or a multi-host configuration, with only the lightweight monitor on the server. The multi-host configuration minimizes any possible server performance impact. Andromedia can support any server vendor with any operating system by porting ARIA server monitor to other platforms while leaving the recorder on a dedicated Solaris machine. The server monitor runs on Solaris, SGI, HP, and NT.

There are two specialized APIs that allow rapid access to the low object classes through a high-level scripting language. The first API allows the reporter access to the information objects in the ARIA database. The second API provides for streaming raw data to the ARIA recorder. This API is the HTTP message protocol and is well defined in the marketplace with ARIA extensions. The importance of these APIs is that extending the product, in terms of data sources (Web sniffers, log files) and output types (e-mail reports, etc.), is greatly simplified.

In addition to internal interfaces, a number of other interfaces are available to users and application developers, including:

- ODBC interface based on DB Connect and Open Access from Object Design, Inc.
- Component interfaces (ActiveX and JavaBeans) for embedding ARIA information in other applications
- A Perl 5 interface for the creation of specialized text-based reports

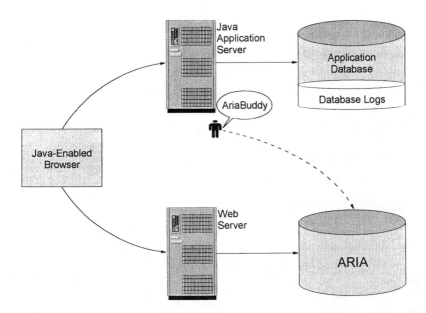

FIGURE 5.11   Logging Java applet activities.

Many CGI, Shockwave, and Java applications communicate with a database or an application that is distinct from the main Web server. This presents several problems for traditional Web activity analysis tools, which rely on server logs to generate reports. A visitor can start and finish an entire application without adding a line to the server log. Consequently, a log file–parsing or packet-reading tool cannot report on the interaction. Java applets are good examples since their prominence on the Web is increasing. Java applets are pieces of code that are downloaded and executed on the client's browser. Once the applet is fully loaded, all interaction occurs through the applet running on the client's browser and there is no communication with the server. Log analysis and packet-reading tools can report on the number of times the applet was downloaded, but cannot report on how visitors interacted with the applet.

Figure 5.11 illustrates a simplified server environment with a Web server and a dedicated application server. The transactions are easily measured while someone is communicating with the main Web server; once the user begins to communicate with the application, the communication protocol is no longer HTTP and server logs are not generated. At the same time, for reporting purposes, not all the interaction is important.

To keep track of the activities, Andromedia uses ARIABuddy, which is a Java-Beans component. It binds itself to user objects in the ARIA database to communicate with the ARIA recorder. Its primary benefit is its ability to send messages that the ARIA recorder understands. Additionally, as a reusable component written in Java, it can plug in directly to a Java application, exist on a Web page as an applet, or be dropped into a Visual Basic application using an ActiveX packager. Using Java's event notification architecture, a Java application can incorporate ARIABuddy to record and report on interaction with the application regardless of where the application is being run — client or server.

ARIA represents openness, flexibility, and performance. The benefits of the product are:

- Combination of different data collection techniques
- Support for a fully automated process
- Support for real-time activity recording
- Open access to processed data
- An object-oriented database
- Real-time reporting despite high traffic volumes

5.6.2 SiteServer from Microsoft

As intranet and Internet Web sites have grown in sophistication, the cost of deploying and maintaining a Web site has risen proportionally. The complexity and the business-critical nature of these sites have also increased the demands on business managers, site developers, and site administrators. As a result, Web sites have become corporate assets that need to be managed, measured, and enhanced to take advantage of new business opportunities.

Companies do not want to change their existing way of doing business to take advantage of intranet technology. Rather, companies want to extend their existing business processes to the Web. SiteServer is a comprehensive intranet server, optimized for Microsoft Windows NT Server and Microsoft Internet Information Server. It enables users to build cost-effective Web solutions for the targeted delivery of information. SiteServer can accomplish the following tasks:

- Efficiently submit, stage, and deploy the latest content while managing and troubleshooting the Web site environment
- Provide site administrators with a tool to gather and index information and create keyword search capabilities on particular sites
- Personalize the user experience by delivering targeted information using dynamic Web pages, direct mail, and personalized push channels
- Improve access to online corporate information and provide subject matter experts with a tool to share their expertise with colleagues
- Analyze site content and site usage to ensure an optimal return on Web site investments

This product is a combination of content authoring and log file analysis. Basically, it supports the publishing, searching, delivery, and analyzing phases. SiteServer provides an application framework for managing business information on the Web. It includes capabilities for adding content to a Web server, organizing information in terms of end users and ad hoc categories, and personalizing the delivery of information through various electronic mechanisms (Figure 5.12).

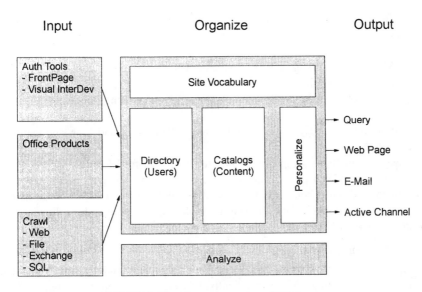

FIGURE 5.12   Key components of SiteServer.

Basic tasks of SiteServer are:

- Publishing content on Web sites: Use the content management and content deployment features to manage and deploy content inside the corporate firewall or over the Internet.
- Creating search capabilities on Web sites: Use the search feature to create keyword search capabilities on Web sites.
- Delivering targeted information over the intranet or Internet: Use personalization and membership, push, and knowledge manager to build a close relationship with site visitors and deliver targeted information.
- Analyzing Web sites: Use the content analyzer feature to analyze the content and structure of Web sites. Users can take advantage of usage import, custom import, and report writer to analyze site usage and ensure an optimal return on Web site investments.

5.6.2.1   Publication Phase

Microsoft provides a set of features for building, staging, and developing Web sites and Web-based applications. SiteServer contains two features that help make publishing content to Web sites reliable and secure:

- Content management
- Content deployment

SiteServer supports the content developer's authoring tool of choice. SiteServer includes:

- Microsoft FrontPage® Web site creation and management tool (see Chapter 4 for more details)
- Microsoft Visual InterDev™ Web development system

These tools make it easy for authors to create Web pages and Web applications, respectively.

SiteServer includes sample code that makes it easy for Webmasters to develop their own custom Web pages. The Starter Sites take this sample code to the next level. Integrated with FrontPage, the Business Internet Site, Employee Information Site, and Online Support Site are easy to deploy and customize by following the FrontPage task list. They are available for download.

5.6.2.1.1   Content management

Content management provides companies with a seamless environment for content sharing and management. More specifically, it is a tool that supports content authors and site administrators. Content authors can use content management to submit, tag, and edit content. Site administrators can use content management to review, manage, and publish content on a corporate intranet or the Internet.

The following tasks can be accomplished:

- Submit content for publication: For content authors, determining where on a corporate intranet to submit content is a fundamental problem. Content management provides content authors with a simple interface to submit and modify content on a server. Content can include documents, images, or Web pages. Using a Web page, content authors can drag and drop their content to a server. During content submission, the content author is asked to tag the content using predefined tags. Tagging is the process of defining the content type and the content attributes. The content type can be documents, spreadsheets, proposals, and so on. Content attributes are the specific properties of the content, including the author's name, the title of the document, or the date of submission. The content type and the content attributes are used by the site administrator to approve and manage the content.
- Perform common content management: Managing documents on a corporate intranet can be a time-consuming process. Site administrators are constantly overloaded with requests for content approval and site reorganization. Content management provides site administrators with an intuitive interface to view, approve, and manage content. Site administrators can manage the approval of content by defining approval settings based on the content type. For example, an administrator can specify that all proposals require approval before being posted to an intranet site. Or, an administrator can specify that all spreadsheets do not require approval before being posted to an intranet site. By defining approval settings,

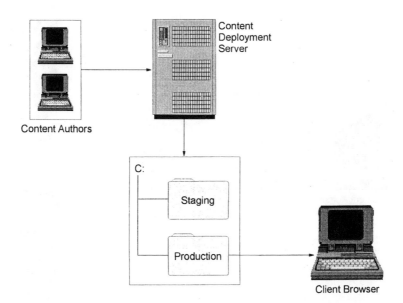

FIGURE 5.13   Content management scenario for small production sites.

administrators can determine whether documents appear immediately on a Web site or after approval from a site administrator. When a document is tagged and submitted, it appears in a specific view of the Web site. If the Web site is in need of reorganization, it can become very expensive to retag each document and reassign the document to a new view. Content management provides administrators with the ability to apply new content attributes to multiple documents. Using this feature, administrators can apply the same content attributes to hundreds of documents and move the documents to a new view within a matter of minutes.

5.6.2.1.2   Content deployment

When content management is complete, site administrators need an efficient method to stage and deploy content across a corporate intranet or the Internet. Security concerns, network outages, and obsolete file transfer protocols have made large file transfers difficult. FTP is no exception.

Content deployment is used for staging and deploying content in Web site environments. With this feature, Webmasters can easily replicate file-based content, such as files, directories, metadata, and access control lists (ACLs) from directory to directory, between local servers, or across the Internet or a corporate intranet. It deploys content to Windows NT and Unix destination servers. This feature can also be used to replicate and install server applications, including Microsoft ActiveX components and Java applets.

Content management is scalable. For small production sites, content deployment features a single server with two distinct roles: content staging and content deployment. With content staging, a site administrator can test or review the content in a

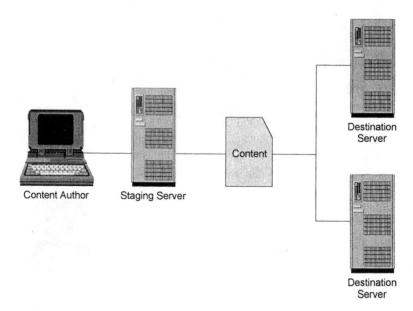

FIGURE 5.14   Content management scenario for large production sites.

staging directory prior to content deployment. When the content is ready for deployment, it can be replicated from the staging directory to a production directory on the same server. When the replication is complete, the content is accessible by client browsers. Figure 5.13 shows a content deployment scenario for small production sites.

For large production sites with multiple servers, each server performs a specific role. The staging server can receive files from content authors. Specifically, content deployment allows the IIS to accept Web content posts from multiple sources by means of a standard HTTP connection. Prior to deployment, the content on the staging server can be reviewed or tested. When the testing is complete, the staging server replicates and deploys the content to multiple destination servers. Once the content is received by the destination servers, the content is placed in a production directory and is accessible by client browsers. Figure 5.14 displays a content deployment scenario for large production sites.

5.6.2.2   Searching Phase

The amount of information available online on the Web is increasing. As a result, site visitors will find it increasingly difficult to find specific information. If site visitors are unable to find the information they need, they will not spend time on particular sites. SiteServer contains a feature called Search, which users can implement to create keyword search capabilities. Keyword search capabilities enable site visitors to use highly targeted search queries to find and retrieve the specific information they need. Search supports the following tasks:

- Gather information: Users can gather and index information by crawling the Internet, and intranet, exchange public folders, newsgroups, an ODBC database, or a file system. For instance, Search can crawl a file system on a local or remote computer to extract the properties and contents of the document. Document properties are the characteristics of the document, including the author's name, date created, title, subject, document's address, and contents of the document. Once extracted, the document properties are stored in a central location, called a catalog. When gathering information, sites to be crawled can be defined. For instance, crawls can be incremental, meaning that users can index only those documents that are new or have changed since the last time the site was crawled. Additionally, users can limit which intranet sites to crawl. The crawler can be restricted to gathering and indexing documents within an intranet and ignore links to the Internet. Also, the notification method can be used to enable Search to gather and index a document. With this method, Search is notified when a document is created or changed and the properties and contents of the document are extracted. When Search crawls a document, the contents and properties of the document are indexed and stored in catalogs on a Build Server. The catalog files are then copied to a Search Server that will begin processing queries for site visitors. For small production sites, Search can build and store catalogs on a single server, as well as begin processing queries for site visitors.

- Initiate a search: Search includes sample pages users can implement to provide site visitors with a Web page from which to initiate a search and a page to display the results of the search. These pages can be extensively tailored to the needs of site visitors and to the information available in the catalogs. When a site visitor initiates a search, any documents in the catalog that match the search criteria are displayed in a Search Results page. Typically, this page displays a list of URLs that match the site visitor's search criteria. When users click a URL, the corresponding document is displayed.

- Improve the quality of catalogs: The quality of catalogs and the quality of the search can be improved by categorizing or tagging content. The tag tool can be used to categorize content, by assigning specific properties to the content. For example, if a financial firm wanted to improve the quality of their catalog, they would use tags to categorize the specific articles that are available in the catalog. Instead of tagging all of their articles as financial articles, the user would tag articles according to their specific content, such as stock articles, bond articles, and mutual fund articles. If users apply specific tags to the documents prior to the documents being cataloged, the granularity of information in the catalogs will be enhanced. When site visitors create a query, they can target their queries to retrieve very specific information.

Search gives users one place to go to find information stored throughout the organization. A robust and powerful enterprise search server enables users to search across a range of intranet and Internet Web sites, files, databases, and exchange public folders.

5.6.2.3   Delivery Phase

In addition to making it easier to find information on the private intranet, SiteServer gives users a number of ways to deliver that information. The following features of SiteServer can be used to deliver targeted information:

- Personalization and membership — It builds a membership community and delivers personalized information.
- Push — It distributes information through channels.
- Knowledge manager — It manages information on Web sites.

5.6.2.3.1   Personalization and membership

It provides information customized to user needs. Active user object (AUO) lets users store properties in the file system or ODBC databases. The rule manager simplifies the building of personalization rules based on user properties and the site vocabulary. Direct mail lets users schedule personalized mailings via SMTP. It organizes and manages membership groups with SiteServer. Using an LDAP directory service and an authentication extension to Microsoft Windows NT operating system security, this feature is ideal for large Internet sites.

The features and functionality of personalization and membership can be categorized as follows:

- Personalization features enable users to offer different content automatically to different users, based on their needs and preferences; users can deliver the content using dynamic Web pages, direct mail, and personalized push channels.
- Membership features enable users to register members and manage member data, control access to site content, and support a sense of community among members.

Fundamental to the operation is the membership directory. It is a central data repository for storing member records, organizational information, and personal profile information. This directory is scalable. For large production sites, the directory can use MS SQL Server as its underlying database. For smaller sites, an access database may be used. Regardless of which database is selected, both will meet the performance expectations. The membership feature supports the following tasks:

- Manage members
- Build a community of members
- Manage billing for purchases, memberships, and subscriptions

For Web sites to be successful, they must attract and retain users. One way to do this is through personalization. This feature can be used to automatically provide different content to different users. Before personalized information is delivered, characteristics, preferences, and behavior of users must be understood. User preferences are stored as attributes. Attributes and other information about users are stored in the membership directory. Information for this directory can be created by users, scripts, or a system administrator. For example, users can record preferences for the type of information they would like to see on particular sites, such as performance reports or status reports.

Information that is not provided by users can be learned from their behavior or acquired from existing databases. The analysis feature of SiteServer may be helpful as well. It can generate user and usage reports. For instance, the usage import feature can gather and import data about how users interact with a particular site. The report writer is helpful in identifying the parts of sites visitors visit most. The reports can be printed and distributed to help Webmasters enhance the content of sites.

Once users preferences are understood, content must be identified that matches their preferences. A content source can be information in individual files, newsgroups, public folders, or open databases.

To approach customers as individuals, Microsoft has developed features into SiteServer that deliver personal, dynamic content by means of a Web page to reflect the preferences and behavior of the user. Personalization sections are areas within the page populated with content that is tailored to the individual user. For example, a Web page can welcome the user by name or display specific content based on personal attributes that are stored in the membership directory. These sections of the Web page are defined using the rule builder. The rule builder is a tool that enables users to tell the system what content to display and when to display it. It uses the attributes stored in the directory to determine what content to deliver. When users are automatically provided with information of interest, they don't spend time navigating particular sites. More important, providing repeat site visitors with the information they need builds a sense of site loyalty that inspires frequent visits.

Direct mail is similar to an electronic version of paper-based direct mail. The primary difference is that direct mail can be used to deliver personalized content to thousands of users in a cost-effective manner. The content of direct mail is determined in much the same way as the personalization section in a Web page. With the attributes and registered user profiles stored in the membership directory, content can be matched with users based on their personal preferences and needs. With direct mailer, a feature of personalization, the mailing can be assembled. As with paper-based direct mail, obtaining the address of the recipient is critical to success. The analysis feature of SiteServer can help define the mailing list. Also, a dynamic mailing list can be generated by running an SQL query against an analysis database.

5.6.2.3.2   Push

Push is the easiest method to create channels for Microsoft Internet Explorer. The active channel server includes channel agents, which allow users to create channels from databases and file systems, as well as searches and indexes. The active channel multicaster dramatically reduces network bandwidth by using multicast technology to deliver channels.

For many users, locating and filtering information is a time-consuming and cost-prohibitive process. Push is the vehicle to automate the process of information delivery. Because users are provided with channels that deliver specific content, they simply subscribe to a channel of interest. The users' browser then periodically requests updates from a Web server, and the information is pushed or delivered to the browser on their desktop.

The active channel server is the primary component of push. It uses the channel definition format (CDF) to deliver information directly to client browsers. Information is delivered through channels. A channel is a vehicle for delivering specific information to a user's desktop. A channel can be either a content or an application channel. A content channel focuses on a particular theme or subject. An application channel can contain program files, controls, or software upgrades for site visitors.

In addition to creating new channels and collecting the channel content, the push feature of SiteServer provides the following channel delivery methods: managed push, multicast delivery, personalized push, and mail delivery.

Managed push — This uses a client's browser to deliver channels to the user. To begin the process, a CDF file must be placed on a Web server. A CDF file contains tags that store information about channels and content items, schedules how often the content updates, and determines how to deliver the content to the user. When a user subscribes to a channel, the user's browser periodically engages the Web server and pulls the project's CDF file to the user's desktop. Managed push uses a delivery schedule specified by the user. For example, users can specify automated refresh, which automatically refreshes the project properties and updates the channels. This ensures that they have the most current information available.

Multicast delivery — Multicast delivery may help to conserve network bandwidth. The active channel server and the SiteServer active channel multicaster service work together to enable the multicast delivery of CDF files. The active channel server builds CDF files, and the SiteServer active channel multicaster service delivers the channels and the content items referenced in the CDF files to self-selected users. Multicast delivery sends a single copy of the data across the network to multiple recipients. The primary benefit of multicast delivery is reduced network traffic, thereby increasing network efficiency.

Personalized push — Push integrates with the personalization feature of Site-Server to provide personalized channels. Personalization takes place at the channel level. Users subscribe only to the channels they want to receive. Thus, users are not overloaded with unwanted information. Once a user subscribes to a specific channel, the personal channel builder dynamically generates a CDF file in response to the user's request, and the channel is delivered to the user's desktop. Additionally, site visitors can use the knowledge manager customizable application to decide which channels they would like to receive.

Mail delivery — Push integrates with the direct mailer feature to deliver channels through e-mail. Unlike other delivery methods, mail delivery does not rely on a CDF file to deliver channels or content items. Instead, channel information, such as links to specify content items, is placed in a mail template. The direct mailer then delivers the e-mail message to users listed in the project's mail distribution list. When the user receives the mail, the channel information is displayed, and he or she can click a link to visit the site of interest.

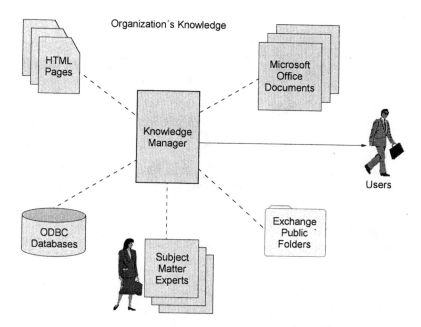

FIGURE 5.15   Use of knowledge manager for information sharing.

The push feature of SiteServer provides content to users without requiring them to browse a specific Web site. Push also gives content providers a tool to ensure the visibility and accessibility of their products, information, and services.

5.6.2.3.3   Knowledge manager

This is an end user application that makes it easy for visitors of the site to find information and receive updates when information is added or changed. The search center, channel center, and briefing center make it easy for visitors to find information they are looking for and have it delivered in the way they wish. Figure 5.15 shows how knowledge manager can improve information sharing.

It leverages several features of SiteServer, including search, personalization and membership, and push, so that users can organize information, manage shared information, and personalize how new information or updates are delivered.

5.6.2.4   Analysis Phase

Determining how users/visitors spend their time allows Webmasters to customize sites for an optimal return on the organization's investment. SiteServer can be used to accomplish the following (Bock, 1998):

- Site usage analysis: Understand how site visitors interact with Web sites
- Site user analysis: Target particular messages to increase the likelihood that site visitors will be interested in the information that sites provide
- Site content analysis: Analyze the content and structure of Web sites

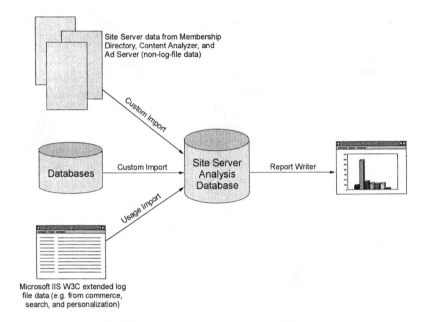

FIGURE 5.16   Reporting feature with SiteServer.

5.6.2.4.1   Site usage analysis

It visualizes the structure of the sites as they are being used, to quickly identify patterns and trends. It identifies high-traffic reports for advertisers. It can import non-Web data to compare usage patterns with other business information. With SiteServer, users are able to analyze data from advanced Web features such as search queries and results.

Understanding how users interact with particular sites is critical to the success of the site. For business managers or site administrators, it can be time consuming and costly to translate network log files into usable information. The analysis feature provides the tools they need to determine how users are spending their time on particular sites. More important, this information can be used for an optimal return on Web site investment.

The features of analysis include: usage import, custom import, and report writer. These features help to analyze Web site usage, produce comprehensive reports, and elicit valuable insights for making informed decisions about Web sites. Every aspect of a site can be analyzed by means of approximately 45 preconfigured reports.

Figure 5.16 illustrates how SiteServer processes complex server log files. In this picture, the usage import feature is used to import server log files and employs several algorithms to reconstruct the actual visits, users, and organizations that interact with the Web site. The information is stored in a database. Then, the report writer queries the database and generates a detailed report based on site activity.

The usage report, custom report, report writer, and scheduler can be used in combination to accomplish the following tasks:

- Manage servers and build a usage database — Usage import provides an organizational structure for easily managing log file content. Customers can utilize usage reports to import and filter log files. For example, each time a user interacts with a particular site, information about the interaction is recorded. The information is imported and filters into a relational database. Usage import examines the data and employs several algorithms to reconstruct the actual visits, users, and organizations that interact with a particular site. Site administrators can then use that information to produce analysis reports such as the one shown in Figure 5.16.
- Design complex analysis — The report writer can be used to view log file data for a single site or multiple sites. By generating reports using the log file data that is imported into the database from the usage import or custom report, administrators can analyze every aspect of the Web site — for example, see how visitors navigate sites, determine which page is viewed most often, and understand the geographic distribution of site visitors. Both canned and custom reports can be generated.
- Integrate data from other SiteServer features — The custom import feature can be used to cross-reference data from multiple Web sites or integrate data from other Microsoft BackOffice servers and SiteServer features, including search and personalization and membership. Once the data is imported, users can launch report writer to generate reports that integrate and analyze the usage data. Custom import can also be used to categorize and classify Web site usage data for enhanced analysis. For example, users can create, view, and edit data properties in the database. Users can also modify the structure of the database.
- Automate analysis — The scheduler feature can be used to automate the task of analyzing and importing log file data. It is ideal for unattended tasks, such as importing log files, resolving IP addresses, and running reports.

Figure 5.17 shows the ten top pages of a site over a period of approximately four weeks. Figure 5.18 displays the ten longest viewed top directories. Figure 5.19 shows the number of requests by hour of the day, which helps in analyzing the daily load on intranets. It is obvious that the busiest hours are between 10:00 a.m. and 4:00 p.m.

5.6.2.4.2　Site user analysis

A set of standard reports helps users understand how the site is being used. Also, customized reports may be created. Reports are presented in familiar formats, including HTML, Microsoft Word, or Excel, and can be generated from more than 25 different Web servers, including IIS, Netscape, Apache, and O'Reilly. SiteServer helps users build stronger relationships with customers by targeting messages to match customer interests. Whether users are selling goods on the Web or providing high-value content, they can use targeted e-mail campaigns to keep customers informed when new versions, new products, special offers, or new content become available that may interest them. The messages will be well received and appreciated.

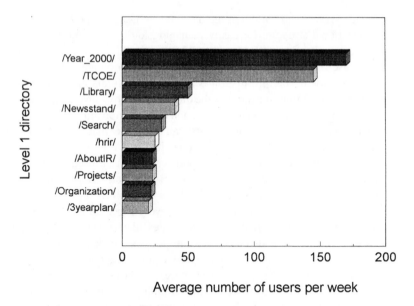

FIGURE 5.17   Top ten pages.

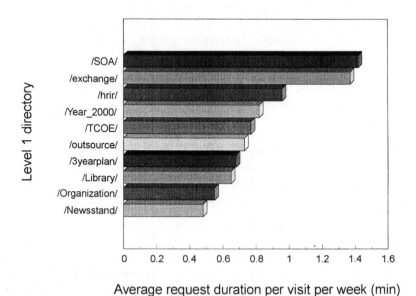

FIGURE 5.18   Ten longest viewed top directories.

Additionally, personalization of content may enhance customers' online experiences. Depending on the event, information can be sent in a particular metropolitan area or distributed nationally. The ability to target communications to customers

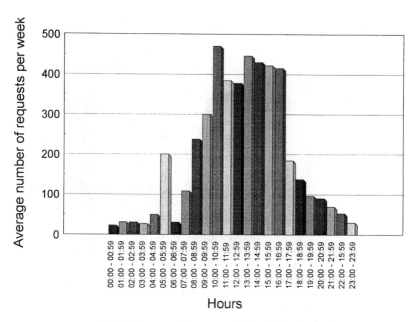

FIGURE 5.19 Requests by hour of the day.

can be considered as a differentiator between sites. The following features can be supported:

- Create user lists for targeted e-mail campaigns: Customers should be associated with target audiences to deliver targeted messages that produce strong responses to actions. Using business rules developed in the rule builder, the analysis feature creates a customer list based on online behavior. For example, the rule builder could request analysis to identify all customers who visited a specific content three times during a week.
- Correlate user interests and contents: Users should drive their content creation process with a strong insight into the interest of their customers. Reports detail top customer interests, as well as correlating their interests with content topics and types.
- Identify user locations: Users can better serve their customers if they know geographic information about customers. This information can be used to locate mirrored sites in heavily trafficked locations for improved customer experience.
- Measure effectiveness of targeting rules: Users are expected to identify which rules in the rule builder are most often used to determine content displayed along with the content that is typically displayed for each rule. This feedback mechanism should be used to fine-tune rules to deliver even better targeted content to customers.

Figure 5.20 analyzes the average visit duration by weekdays. The figure shows that the daily visit duration is stable.

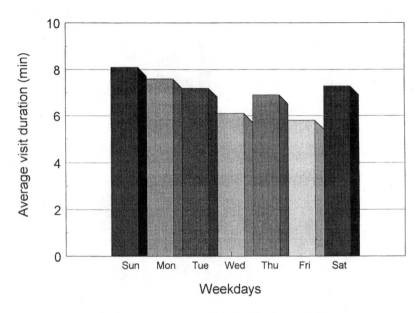

FIGURE 5.20   Average visit duration by weekdays.

5.6.2.4.3   Site content analysis

Monitoring complex Web sites can be a time-consuming task. Links, site content, and site resources can be difficult to track down in a timely manner. Using the task-oriented approach provided by the content analyzer, users can analyze the content of sites quickly and efficiently.

Content analyzer, a feature of SiteServer analysis, lets users execute Web site management tasks from a central location. Content analyzer is designed to show the content of particular sites at a glance. With a map as a starting point, site structures can be analyzed, contents tracked, local and remote sites maintained, and access to key content improved. Content analyzer provides more than 20 analysis reports.

The following tasks can be performed by the content analyzer:

- Visualize the site — Content analyzer provides site visualization capabil-ities. Using the outline pane or the hyperbolic pane, users can display the structure and content of particular sites. The outline pane displays pages and resources in a hierarchical tree. Users can expand or collapse the tree to display dependencies or interrelated site resources. The hyperbolic pane displays a map of the site. A site map is a visual representation of the site. Site maps reveal how the structure, content, and resources, such as images, audio files, and video files, are linked to each other. Maps can also be used to show what has changed on a particular site. Changes, such as new or deleted pages, new links, etc., can be identified by comparing an old map to a new map of the site. By comparing maps, users can see what is new, what has changed, and what no longer exists.

- Analyze site content — In order to streamline Web site management, content analyzer can provide detailed information based on site reports or custom searches. This information can be used to monitor site content and maintain site quality. Searches can be broad or narrow, can be for outdated content, can focus on a single resource type, or may focus on multiple resource types. All site views, reports, and search results can be printed and distributed to help Webmasters solve problems and enhance the usability of particular Web sites.
- Improve access to key content — Content analyzer can verify the status of offsite links and ensure that site visitors are provided with the most direct route to key content. Routes may be identified based on usage data. Most sites have multiple routes or links to specific content. Content analyzer can examine all the routes that point to a specific resource and identify the most frequently traveled route. If the most frequently traveled route is not a direct route, content analyzer can show how to reorganize the site to improve access to popular content.

Content analysis generates site maps and content reports to examine site integrity and consistency. It identifies broken and external links, duplicate and orphan pages, and pages with large load sizes. Webmasters are concerned about users' experiences while visiting particular sites. They are also concerned about the site's conformance to company policy and the efficient use of site resources. A good user experience is a function of not only valuable content, but also easy navigability, well-positioned content, fresh content, and a uniform look and feel throughout the site. In monitoring conformance to site policy, Webmasters need to verify the address of offsite links, confirm accuracy of e-mail addresses listed on the site, and ensure proper use of copyright statements. Lastly, Webmasters should identify and remove duplicate content and pages that are no longer referenced (orphans) to save server capacity. Content analysis offers the following:

- Uncover valuable content and site information
  Web maps: SiteServer analysis maps Web sites by extracting vital content attribute information. Analysis uses a "spider" to record the links on a page and then follow those links to locate new pages and new links, storing all the information in a Web map. Each map is an information base that provides a catalog of the Web objects in a site (HTML pages, Java and PDF files, graphics, links, and others) and properties associated with those objects. Once created, a Web map provides a visual representation of a site that can be used for navigation, site analysis, and site maintenance.
  Partial mapping: When creating a Web map, users can map all or part of the site. They can map just the portions of the site that interest them. The parts of the site not explored are listed in an "unexplored" report. Users can also build maps that have information from multiple sites. With analysis, users can control exactly what resources are and are not included.

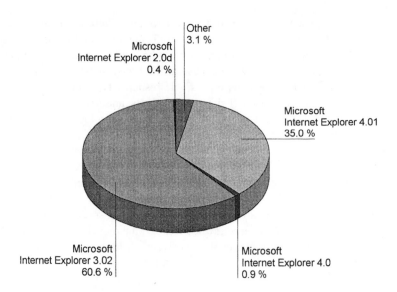

FIGURE 5.21   Distribution of browsers.

Mapping through proxies, secure sites, cookies, and forms: If sites have secure portions or a secure proxy, users can still take full advantage of analysis by first specifying user names and passwords. This capability can also be used to test the authentication settings of sites. Additionally, support for mapping sites, requesting cookies, and requiring form completion is available.

Scheduled Web site updates: This ensures that Web maps are updated when users need them.

Internet-friendly spider: The analysis spider minimizes the effects of its activity on the sites that it maps. The spider does not attempt queries, avoids looping and repetition, and does not access any page more than once, even if that page is pointed to from several other locations. The spider can be paused or aborted by the user at any time during the mapping process, and the mapping functions of the spider come with default settings that do not overload Web servers.

- Visualize site structure

Hyperbolic view: The hyperbolic view gives users a dynamic overview of the site that clarifies relationships and interconnections and helps users immediately grasp the overall structure of a site. Details of any part of the map can be zoomed. The outline and hyperbolic views are linked, so that the object of focus remains selected in either view.

Outline view: The outline view displays the site's object hierarchy, giving a familiar, detailed view of the files in a site. Users have complete control over which tree hierarchy is displayed.

Browser view: Users can see what a page looks like on the Web, or they can view its HTML source. Figure 5.21 shows the distribution of browsers used to access a particular site.

Filtered views: Users can choose to hide or show particular types of resources, such as images or gateways, and further customize the views to show only what users really want to see.

- Report on content and site quality and consistency

Predefined and custom reporting: Analysis provides users with detailed content information based on site reports or custom searches. This information can be used to monitor content and maintain site consistency and quality. All site views, reports, and search results can be printed and distributed to help Web teams solve problems and develop sites that are easier to manage and navigate. Users can select exactly what properties are shown on customized reports (there are approximately 70 properties to select from), and in exactly what order. These reports can be printed, exported to spreadsheets or databases, or displayed as HTML tables in browsers.

Site summary report: This report gives a detailed overview of the site including object statistics, a status summary, map statistics, and server information. From this report, users can forward to other detailed reports for each aspect of the particular site and the objects in it.

Detailed resource reports: Users can get detailed information about individual pages or images. They can find out what mail addresses and newsgroups they are using, as well as what files are being made available for FTP download.

Site integrity reports: Each duplicate object is shown with its full URL and an icon indicating the type of object in the "duplicates" report. Site resources that no longer have references to them, or orphan files, are listed in the "unreferenced objects" report.

Site comparison and update reports: The comparison summary, what is new, and what has changed reports enable users to compare Web maps to see what has changed and to get detailed information about the changes.

Table of contents: Users can see the entire site structure in hyperlinked HTML format with the hierarchy report, which shows the basic structure of the particular site in a tree-like, outline form. Each resource is shown with its URL and a graphical representation. This report information can be used to quickly generate and publish items that enhance site usability, such as a fully customizable table of contents and linked indexes.

Remote reporting: Users can view and run content analysis reports from any universal browser. Users can also schedule their favorite reports to be ready every day with the latest information about the site.

META tag support: Users can incorporate META tags in a Web map using the SiteServer tag tool, making it easy for content authors to insert their names and editing date right on the page. They can then search and sort on the META tag data and quickly identify a page's author and when the page was last changed.

Independent authoring tool: When users want to make changes to the content, analysis integrates seamlessly with text, audio, video, and program-editing software. Analysis launches the appropriate helper application with just a click on a map object.

- Search for problem pages

  Resource searching: Powerful search capabilities let users locate any information contained in particular sites by searching for resources such as text strings, author, expiration date, or other user-defined parameters. The results of these queries can be sorted, printed, exported to HTML, and exported as tab-delimited text for importing into spreadsheets or databases for further analysis.

  Quick searches: Analysis gives users easy access to a set of common searches including pages with load size over specific thresholds, top entry pages, and pages with external referrers.

  Search scope: Users can search the entire map or only what's currently displayed. They may also search for a specific object type or property such as its size or URL, or a text string.

  Distributing search results to the Web team: A built-in e-mail interface allows users to easily send the results from quick and custom searches to the appropriate members of the Web team to take action. Analysis manages Web site maintenance and improvement efforts by creating checklists of site fixes such as out-of-date pages, duplicates, orphans, images without ALT tags, and others.

5.7   SUMMARY

Log file analysis is the key to obtain detailed data about site usage metrics. Most products referenced here are rich on functionality in terms of providing answers about users, site usage, and content statistics. Reports are available on Web servers to be accessed by universal browsers. In the case of SiteServer from Microsoft, the first signs of larger scale integration can be observed. SiteServer embeds FrontPage and is able to work with SMS (systems management server). Also with other tools, the integration with management platforms and frameworks is an obvious target.

6 Traffic Measurements

CONTENTS

6.1 INTRODUCTION

Log files are not the only source of information for analyzing Web sites. Other tools reside "on the wire" or LANs and collect information on performance and traffic metrics. The information depth and the overhead are significant indicators that may differentiate between log file analyzers and these products. In certain environments, the most effective results can be achieved only when both types of tools are deployed in combination.

Over the past several years, companies have adopted distributed multi-tier network infrastructures and moved business operations from traditional client/server applications to distributed Web-based applications. However, as more and more users come to depend on Web servers and TCP-based services, IT organizations are discovering that their current infrastructures are unable to offer the performance and availability expected by users; nor do they provide the management and monitoring capabilities required by IT organizations themselves.

Over the past several years, large corporations have begun reengineering their enterprise networks and establishing distributed, multi-tier infrastructures. These multi-tier infrastructures typically include three levels:

- The wide area network (WAN) level enables communication across multiple points of presence (POPs).
- The Web level supports server farms providing a wide range of TCP-based services, including HTTP, FTP, SMTP, and telnet.
- The application level supports farms of application servers that offload computation from Web servers to increase overall site performance.

IT organizations are currently deploying new distributed, Web-based applications to take advantage of this new enterprise infrastructure. In place of fat software clients and centralized application servers, corporations are deploying Web browsers on every desktop, Web servers in departments and divisions, and application servers at multiple locations.

The new Web-centric model offers several advantages over the client/server model it replaces. IT departments can deploy Web browsers quickly and affordably to every desktop platform. Basic Web skills can be learned quickly and are popular with users. If an application requires modification to reflect changing business practices, IT departments need only modify the application itself, not the complex clients that used to work with the application. Most important, distributed, Web-based infrastructures move content and applications closer to users and provide improved reliability and availability. Employees can leverage this new infrastructure to improve internal business practices, communication with partners and suppliers, and services for customers.

Although distributed, multi-tier infrastructures offer considerable advantages over earlier network architectures, they still do not offer the performance and availability expected by end users; nor do they provide the management and monitoring capabilities expected by IT organizations. Multi-tier architectures are physically well connected, but not logically well connected. Standard network equipment enables traffic to flow, but not necessarily to the server best suited to respond. IT departments deploying these networks need traffic management solutions that intelligently direct TCP traffic to optimal resources at each tier of the enterprise infrastructure. An optimal traffic management solution requires communication between tiers. For example, there is little point in a DNS server directing traffic to a local server that is down or overloaded while another server is available with processing cycles to spare. To perform its job optimally, the DNS server needs availability and load information from the servers to which it directs requests.

The multi-tier model itself, when implemented with the standard software products available today, does not monitor services for system failures or spikes. Nor

TABLE 6.1
RMON MIB Groups for Ethernet

Statistics group	Features a table that tracks about 20 different characteristics of traffic on the Ethernet LAN segment, including total octets and packets, oversized packets and errors
History group	Allows a manager to establish the frequency and duration of traffic-observation intervals, called "buckets"; the agent can then record the characteristics of traffic according to these bucket intervals
Alarm group	Permits the user to establish the criteria and thresholds that will prompt the agent to issue alarms
Host group	Organizes traffic statistics by each LAN node, based on time intervals set by the manager
HostTopN group	Allows the user to set up ordered lists and reports based on the highest statistics generated via the host group
Matrix group	Maintains two tables of traffic statistics based on pairs of communicating nodes; one is organized by sending node addresses, the other by receiving node addresses
Filter group	Allows a manager to define, by channel, particular characteristics of packets; a filter might instruct the agent, for example, to record packets with a value that indicates they contain DECnet messages
Packet capture group	Works with the filter group and lets the manager specify the memory resources to be used for recording packets that meet the filter criteria
Event group	Allows the manager to specify a set of parameters or conditions to be observed by the agent; whenever these parameters or conditions occur, the agent will record an event into a log

does it provide other capabilities that IT departments require to manage busy, distributed networks effectively. Specifically, it provides no:

- Policies for scheduling TCP traffic based on specific centralized events
- Remote management reporting integration with standard network management tools

IT organizations need integrated software systems that can be layered on top of the existing infrastructure to provide intelligent scheduling of requests and information.

6.2 PRINCIPLES OF DATA COLLECTION

To support standardized measurements in LANs, RMON (remote monitoring) has been introduced. RMON can be defined as the extension of MIBs (management information base) supported by SNMP. RMON defines the most important indicators for both Ethernet and Token Ring. These definition are shown in Tables 6.1 and 6.2, respectively. For FDDI, in most cases Token Ring indicators are used.

TABLE 6.2
RMON MIB Groups for Token Ring

Statistics group	Includes packets, octets, broadcasts, dropped packets, soft errors, and packet distribution statistics; statistics are at two levels: MAC for the protocol level and LLC statistics to measure traffic flow
History group	Long-term historical data for segment trend analysis; histories include both MAC and LLC statistics
Host group	Collects information on each host discovered on the segment
HostTopN group	Provides sorted statistics that allow reduction of network overhead by looking only at the most active hosts on each segment
Matrix group	Reports on traffic errors between any host pair for correlating conversations on the most active nodes
Ring station group	Collects general ring information and specific information for each station; general information includes: ring state (normal, beacon, claim token, purge), active monitor, and number of active stations; ring station information includes a variety of error counters, station status, insertion time, and last enter/exit time
Ring station order	Maps station MAC addresses to their order in the ring
Source routing statistics	In source-routing bridges, information is provided on the number frames and octets transmitted to and from the local ring; other data includes broadcasts per route and frame counter per hop
Alarm group	Reports changes in network characteristics based on thresholds for any or all MIBs, allowing RMON to be used as a proactive tool
Event group	Logs events on the basis of thresholds; events may be used to initiate functions such as data capture or instance counts to isolate specific segments of the network
Filter group	Definitions of packet matches for selective information capture, including logical operations (AND, OR, NOT), so network events can be specified for data capture, alarms, and statistics
Packet capture group	Stores packets that match filtering specifications

Using RMON, there is a common denominator for measuring, processing, and reporting on performance indicators for LANs and interconnected LANs. Intranets heavily use LANs, usually combined with server farms. RMON metrics offer the opportunity to get insight into traffic behavior of users, visitors, and Web applications. RMON-based products can be used the same way as traffic monitors are used in WANs and LANs. They are complementary.

The principal objectives of using RMON standards are:

- Loosely coupling hardware and software probes and the management platform
- Proactive monitoring using probes (monitors)
- Powerful problem diagnosis based on detailed data
- Powerful reporting to support performance analysis and capacity planning

Communication Layers	RMON
Application Layer	RMON2
Presentation Layer	RMON2
Session Layer	RMON2
Transport Layer	RMON2
Network Layer	RMON2
Data Link Layer	RMON1
Physical Layer	RMON1

FIGURE 6.1 Layer model for RMON1 and RMON2.

- Support of multiple management stations by hardware and software probes
- Support of local processing and display functions by the probes

There is a difference between RMON1 and RMON2 standards. RMON1 is best known for monitoring MAC-level (media access unit) level performance within LAN segments. This base, combined with product-specific extensions, can provide valuable information for fault, throughput, utilization, and communication metrics. Using RMON2, networked LANs can also be measured. RMON2 concentrates on the networks and application layers of the communication. The result is that performance analysis can go beyond routers in IP networks. Figure 6.1 shows the different application areas of RMON1 and RMON2. Table 6.3 summarizes the principal indicators for RMON2.

There are two components in implementing RMON-based products. The client consists of a hardware- or software-based probe, implemented in the LAN segments, and of a server that is responsible for centrally processing data captured by the probes (or monitors). Usually, the server is implemented into management platforms. Clients and the server communicate with each other using the SNMP (Huntington-Lee, 1996).

RMON offers multiple benefits:

- Reduction of lead time to solve network-related problems: Probes or monitors can be preprogrammed for multiple problem indicators. If conditions are met, and thresholds are exceeded, traps are immediately sent to the management station. Trouble tickets are opened and referred to the right problem resolution tier.
- Reduction of outage time: Actions can be implemented earlier than ever before, because information is available in real time. Engineers and technicians do not need to be on site to diagnose problems.
- Reduction of traveling expenses: With portable LAN analyzers, engineers and technicians are expected to be on site. With RMON probes, that is

**TABLE 6.3
RMON2 MIB Groups**

Address mapping	Matches each network address with a specific port to which the hosts are attached; identifies traffic-generating nodes/hosts by MAC, Token Ring, or Ethernet address; helps identify specific patterns of network traffic useful in node discovery and network topology configurations; the address translation feature adds duplicate IP address detection, resolving a common trouble spot with network routers and virtual LANs
Network-layer host table	Tracks packets, errors, and bytes for each host according to a network-layer protocol; permits decoding of packets based on their network-layer address, in essence permitting network managers to look beyond the router at each of the hosts configured on the network
Network-layer matrix table	Tracks the number of packets sent between a pair of hosts by network-layer protocol; the network manager can identify network problems quicker, showing the protocol-specific traffic between communicating pairs of systems
Application-layer host table	Tracks packets, errors, and bytes by host on an application-specific basis (e.g., Lotus Notes, e-mail, Web, etc.); can be used by network managers to charge users on the basis of how much network bandwidth was used by their applications
Application-layer matrix table	Tracks packet activity between pairs of hosts by application (e.g., pairs of hosts exchanging internet information)
Probe configuration	Defines standard parameters for remotely configuring probes — parameters, such as network address, SNMP error trap destinations, modem communications with probes, serial line information, and downloading of data to probes; provides enhanced interoperability between probes by specifying standard parameters for operations, permitting one vendor's RMON application to remotely configure another vendor's RMON probe
User history collection group	Polls, filters, and stores statistics based on user-defined variables, creating a log of the data for use as a historical tracking tool (this RMON2 tool is in contrast to RMON1, where historical data is gathered on a predefined set of statistics)

no longer required. This is significant when a large number of LAN segments must be supervised by a centrally located support staff.

- Better scheduling of working time for engineers and technicians: With a centralized work force management system, supported by RMON probes, personnel can be better utilized. In addition, more challenging tasks can be assigned; routine data collection tasks are handled by probes and monitors.
- Availability of history data: RMON probes provide information to be maintained centrally. After processing and compression, performance data can be maintained in databases. These data are available for future performance studies and capacity planning.

The disadvantages of RMON probes must also be considered:

- RMON usage assumes the use of TCP/IP: Most RMON-based products are based on TCP/IP. In environments with proprietary protocols and applications, such as SNA, DSA, DNA, and Novell, probes must be customized to these protocols. Protocol converters introduce additional components and thus additional complexity into the networks.
- RMON probes are expected to be installed in each LAN segment: Continuous measurements guarantee the best results. If probes are rotated, measurement results might not be representative. With RMON2, all limitations of RMON1 can be eliminated. If switches are heavily used in networking structures, even more RMON probes are required.
- Costs of probes can be a problem: The price/performance ratio of probes is improving constantly. The price seems to be reasonable for individual probes, but large volumes require heavy investments. Most probes can measure multiple segments — usually up to four — further improving the price/performance ratio. In the case of many single remote segments, additional investments are required.

RMON probes can be implemented in three different ways:

1. As a stand-alone monitor (Figure 6.2)
2. As a module of hubs, routers, and switches (Figure 6.3)
3. As a software module in Unix, NT operating systems, or PC workstations (Figure 6.4)

Each of these alternatives has benefits and disadvantages.

Benefits of probe as a stand-alone monitor:
- Excellent performance
- Support of all functions
- Availability in various options, such as stackable or rack-mountable

Disadvantages:
- High costs for an average LAN segment
- Multiple probes are required, when segmentation of LANs is deployed by switches without using probes or ports
- Most advanced LAN technologies might not be supported right away

Benefits of probe as a module of hubs, routers, and switches:
- It represents a very convenient solution because the networking components have been deployed
- The costs are much lower than in the case of stand-alone probes
- The integration of probes into switches is less expensive than deploying an individual probe in each switched segment

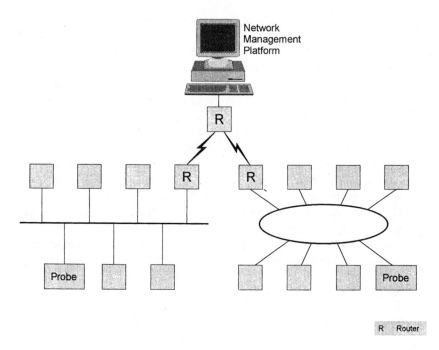

FIGURE 6.2   Probe as a stand-alone monitor.

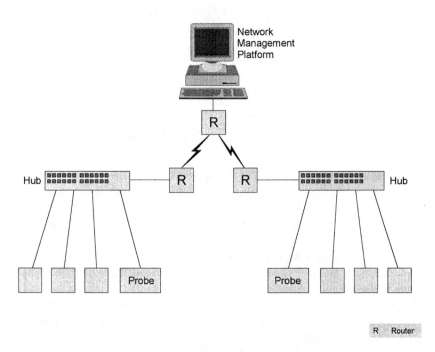

FIGURE 6.3   Probe as plug-in hardware module.

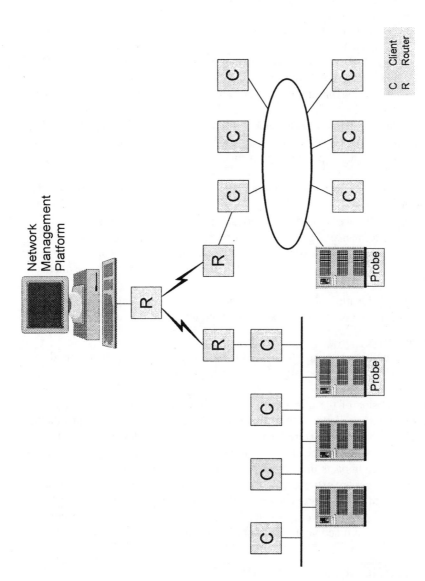

FIGURE 6.4 Probe as software module in Unix or PC workstations.

Disadvantages:

- The networking components need an upgrade or customization to support the probes
- Upgrades are not always economical
- Performance might become a problem
- Conformance to standards may not be met
- Problem with the probe may impact the performance of the networking component
- The RMON functionality may be very much limited; not all RMON indicator groups are supported
- Processing programs are very simple in comparison to stand alone probes
- Integration with management platforms is usually incomplete
- RMON modules may be provided by different vendors, which may lead to incompatibility problems

Benefits of probe as a software module in Unix, NT operating systems, or PC workstations:

- Much lower costs in comparison to the other alternatives
- Performance is good when run on RISC or Pentium processors
- Scalability and extendibility are excellent
- Support of state-of-the-art technology is easier than with the alternatives
- Combination of supervising Ethernet and Token Ring is possible
- Outband access to probes is possible with proper configuration

Disadvantages:

- Purchase of adapters and additional workstations may be required
- The user is responsible for the installation

The extension of the third alternative might also be used for switched LANs. RMON may be installed into the adapter of end user devices. Usually, just the filter and packet capture groups are supported at the end user device level. The other groups are supported by the collector (Figure 6.5). The overhead is minimal, as are the performance impacts in the switch and in the end user devices.

RMON probes are extremely helpful for collecting data on Web site accesses and activities. In order to support interoperability between probes, suppliers of probes and monitors are expected to work together. Standards are continuously being improved, offering even more functionality for capturing and processing Web site–relevant data.

6.3 TRAFFIC MONITORING ISSUES

The targeted metrics are the same as with log file analyzers (Chapter 5), but the source of data is different. When selecting products, there are a number of criteria, such as information depth, overhead, and reporting capabilities, that must be carefully evaluated. The market potentials are good, addressed by few vendors. These

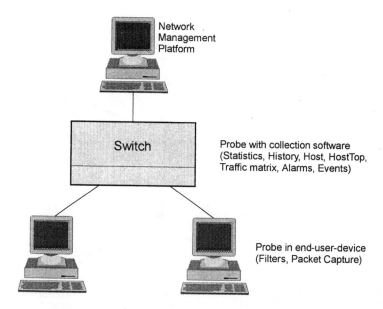

FIGURE 6.5   Probe as software plug-in in end user devices.

criteria are also important when Webmasters want to position traffic measurements within their IT administration or when they want to deploy this functionality within their organization.

The architecture of a product answers the question of whether it can support a distributed architecture. Distribution may mean that collecting, processing, reporting, and distributing data can be supported in various processors and at different locations. Figure 6.6 shows these functions with a distributed solution.

The monitors passively measure the traffic in the network segments. They are actually microcomputers with ever-increasing intelligence. Their operating systems are either proprietary or based on UNIX, or more likely on NT. Usually, they are programmed to interpret many protocols. TCP/IP, UDP/IP, and HTTP are high on the priority list of vendors.

The data capturing technique is essential with traffic measurement tools. The measurement probes are attached to the digital interface of the communication channels. They can reside directly on the network (stand-alone probes) or can be co-located with networking equipment. In this case, the probe is used as a plug-in. Even software probes can be used and implemented into networking components or end user devices. The hardware or software probes usually include event scheduling. This means determining polling cycles and time periods when downloading of measurement data is intended. Transmission should be scheduled for low-traffic periods. Probes are expected to deal with large data volumes. These volumes depend, to a large degree, on visitors' traffic in networking segments. Probes have limited storage capabilities; implementation examples show capabilities up to 24 hours. When this limit is exceeded, measurement data are overwritten by new data. Usually, measurement data are uploaded for further processing. It is important to know how

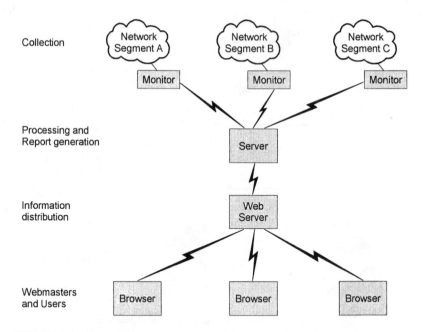

FIGURE 6.6   Generic product architecture for processing traffic measurement data.

uploads are organized and how rapidly they can be executed. As indicated in Figure 6.6, some WANs may show bandwidth limitations. The bandwidth is usually shared with other applications, potentially causing traffic congestion. Bandwidth-on-demand solutions are rare with measurement probes. When transmission is arranged for low-traffic periods, the actuality of measurement results may suffer. In such cases, local storage requirements increase and processing, report generation, and information distribution are delayed by several hours or even by days.

   Two solutions may help. The first is using intelligent filtering during and shortly after data collection. Redundant data are removed from captured packets during collection. Data volumes decrease, local storage requirements decrease as well, but processing requirements of the probes increase. The second solution may use data compression or data compaction with the same results and impacts as can be observed with the first solution.

   Overhead is a very critical issue with large data volumes. Data capturing is expected not to introduce any overhead in the case of hardware-based probes. Overhead is minimal with software-based probes. It is assumed that measurement data are stored away immediately after collection. If local processing is taking place, overhead must be critically quantified. If resource demand is high, probes must be upgraded properly. Data transmission overhead can be heavy if everything is transmitted to the site where processing takes place. Dedicated bandwidth would be too expensive for measurement and management purposes only. If bandwidth is shared with other applications, priorities must be set higher for business applications than for transmitting measurement data.

It is critical that the following data are captured to conduct a detailed Web site analysis of visitors or groups of visitors:

- Who is the visitor?
- What is the purpose of the visit?
- Where is the visitor coming from?
- When has the visit taken place?
- What keywords have brought the visitor to the site?
- What search machines helped the visitor access the site?
- How long was the visit?

Data losses cannot be completely avoided. Probes, monitors, networking devices, user workstations, or transmission equipment may fail; in such cases, there will be gaps in the sequence of events. Backup capabilities may be investigated, but IT budgets won't usually allow administrators to spend too much for backing up large volumes of measurement data. In the worst case, certain time windows are missing in reporting and statistics. Those gaps may be filled with extrapolated data.

Because of the considerable data volumes, databases should be considered to maintain raw and/or processed data. Database managers would then offer a number of built-in features to maintain data. Clustering visitors may be deployed from various perspectives, such as geography, common applications, common interests on home pages, and date and time of visits. ODBC (open database connectivity) support assists the exchange of data between different databases and correlation of data from various databases. Besides measurement data, other data sources can also be maintained in the same data warehouse. Besides routine measurement data analysis with concrete targeted reports, special analysis may also occasionally be conducted. This special analysis, called data mining, can uncover traffic patterns and user/visitor behavior. Both are important in sizing systems and networking resources.

One of the most important questions is how measurement data analysis performs when data volumes increase. A volume increase can be caused by offering more pages on more Web servers, more visitors, longer visits, and extensive use of page links. In any case, collection and processing capabilities must be estimated and quantified prior to deciding on procedures and products.

To reduce processing and transmission load of measurement data, redundant data should be filtered out as near as possible to the data capturing locations. Filters can help to avoid storing redundant data. Filters can also be very useful in the report generation process. Again, unnecessary data must not be processed for reports. Powerful filters help to streamline reporting.

Not everything can be automated with analyzing measurement data. The user interface is still one of the most important selection criteria for products. Graphical user interfaces are the most common, but simple products still work with textual interfaces. When measurement data are integrated with management platforms, this request is automatically met by management platforms.

Reporting is the tool to distribute the results of measurement data analysis. Predefined reports and report elements as well as templates help to speed up the report design and generation process. Periodic reports can be automatically generated and distributed for both single Web servers and Web server farms. In the case of many Web servers, report generation must be carefully synchronized and scheduled. Flexible formatting helps to customize reports to special user needs.

Output alternatives of reports are many. The most frequently used solutions include Word, Excel, HTML, and ASCII. Report distribution also offers multiple choices:

- Reports may be stored on Web servers to be accessed by authorized users who are equipped with universal browsers.
- Reports can be uploaded into special servers or even pushed to selected users.
- Reports may be distributed as attachments to e-mail messages.
- Reports can be generated at remote sites; this alternative may save bandwidth when preprocessed data instead of completely formatted reports are sent to certain remote locations.

Documentation may take various forms. For immediate answers, an integrated online manual would be very helpful. Paper-based manuals are still useful for detailed answers and analysis; however, they will soon be replaced by Web-based documentation systems. In critical cases, hotlines can help with operational problems.

Measurement data analysis is actually another management application. If a management platform is used, this application can be integrated into it. There are many ways to integrate; most likely a command line interface (CLI) will be deployed. Table 6.4 summarizes these evaluation criteria for traffic measurement tools.

6.4   TRAFFIC MONITORS

Information about the use of Web pages, their users, the frequency of access, resource utilization, and traffic volumes can also be collected in the network or at the interfaces of the network. In many cases, the borders between tools and techniques in the server and networking segments are not clear. This part of the chapter will concentrate on several tools. Tools differ with regard to data collection technologies, performance metrics used, and reports offered.

6.4.1   WebSniffer from Network Associates

WebSniffer combines Expert Sniffer Network Analyzer technology with Web performance management tools to create an early warning system of Web performance degradations. WebSniffer differentiates between problems rooted on the network and those occurring on the server.

WebSniffer is a performance management system that analyzes network protocol packets and host operation to quickly identify problems related to end users. It

TABLE 6.4
Evaluation Criteria for Traffic Measurement Tools

Architecture of the product
System requirements
  Hardware
  Operating system
Data capturing techniques
  Locations of measurement probes
  Event scheduling
  Storing measurement data
  Download of processing
Overhead
  Data capturing
  Data processing
  Data transmission
What indicators (metrics) can be measured?
Components of Web site analysis
  Who
  Why
  Where
  When
  What (e.g., keywords)
  How accessed (e.g., what search machine)
  How long accessed
Precautions against data losses
Database management
  Use of a database manager
  Grouping capabilities
  ODBC support
  Data warehouse
  Drill-down (mining) capabilities
Scalability
Flexibility
Filtering
  Reports
  Information within reports
User interface
  Textual interface
  Graphical interface
  Integration with management platforms
Reporting
  Predefined reports
  Use of templates
  Predefined report elements
  Trend reporting
  Use of a report scheduler

TABLE 6.4 (continued)
Evaluation Criteria for Traffic Measurement Tools

Reporting (continued)
    E-mail reports
    Report output alternatives (Word, Excel, HTML, ASCII)
    Flexible formatting
    Cross-tabs by correlating fields in the database
    Customization capabilities
Distribution of reports
    Accessing reports
    Uploading reports
    Remote reporting
Integration with management platforms
Documentation and support
    Integrated online manual
    Paper manual
    Help desk
Vendor
    Number of clients
    Founded when
    What other products are offered
    Support (e.g., on-site, hotline)
    Maintenance contracts
    Financial strength
    Keeps current with servers software releases

---

assesses Web site availability, response time, and user abort rates to isolate whether problems are rooted on the Web server host or are network and client based. For host-based problems, WebSniffer correlates availability, response time, and end user abort rates with host resource constraints to pinpoint Web site problems originating at the host CPU, memory, or disk drives.

WebSniffer monitors, analyzes, and automatically identifies problems with Web site end user experience, all in real time. Trouble conditions trigger alarms both within WebSniffer's browser-based user interface and externally. For Web site managers, that translates into powerful benefits:

- Faster problem resolution — It reduces the time spent troubleshooting and solving problems by automatically identifying problems, suggesting solutions, and providing the data underlying its decision making.
- Early warning of problems — Analysis and alarming occurs in real time, and alarms can be sent outside WebSniffer in an e-mail message, to a pager, via an SNMP trap, or through user-written scripts. This provides immediate notification of problems, even if you are away from the site.
- Remote accessibility — It finds out what is happening wherever the user is. Its user interface can be brought up rapidly in a Java-enabled browser.

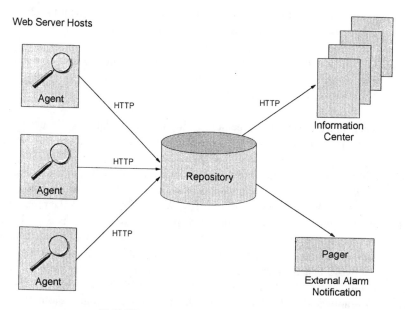

Web Server Hosts

FIGURE 6.7 Architecture of WebSniffer.

6.4.1.1 Components of WebSniffer

WebSniffer has three principal components that gather, analyze, and then display information (Figure 6.7):

- WebSniffer agents are lightweight processes that reside on each Web server host and gather Web site performance data by watching network traffic and host resources.
- The WebSniffer repository analyzes the raw data gathered by the agents and delivers information vital for troubleshooting and managing the Web site. The repository resides on a dedicated host machine.
- The WebSniffer information center displays the information produced by the repository and is the primary point of interaction with the product. It is accessible from Web browsers that support Java.

6.4.1.2 WebSniffer Operations

Having an agent on each Web server host enables WebSniffer to gather data critical to understanding how a Web site is behaving. WebSniffer is able to measure important network and server timing information and characterize host operations.

Web servers and Web browsers communicate via HTTP, which is part of the TCP/IP protocol family. WebSniffer agents observe the TCP/IP protocol stack on each Web server to get a view into the client/server communication process. Collecting detailed timing data for HTTP communications is how WebSniffer derives its end-to-end response time measurements for URLs. When a client wants to exchange information with a Web server it issues an HTTP command such as GET or POST to the server. This process has three phases:

- The client establishes a connection to the server and issues a command.
- The server works to fulfill the command.
- The connection is closed.

During each phase, WebSniffer gathers a unique set of data that is used by the WebSniffer repository to analyze Web site availability, response time, and user abort rates.

WebSniffer agents gather host performance data directly from the operating system API, available from commands such as vmstat and netstat on Unix, or the permon utility on Windows NT. Additionally, WebSniffer accesses data that operating system commands do not consider or display, and WebSniffer is more efficient than these commands in many cases. Regardless of the host data origins, it is what WebSniffer does with it that makes the information it produces valuable. It feeds the raw network and host data to the repository where it is analyzed, and expert rules are applied to extract useful information. WebSniffer then displays this information in graphs and tables for easy visualization of patterns and anomalies.

WebSniffer's Internet agent enables a Web site manager to see how the site is performing from the perspective of users at various spots throughout the Internet and compares it to the responsiveness of other major Web sites. WebSniffer accesses a database of measurements taken by Keynote Systems every 15 minutes from ten major U.S., Asian, and European cities. The repository polls the Keynote server, downloads response time information for a specified URL, and compares that data with 40 internet performance indices. This index averages the response time for accessing and downloading the home pages of 40 Web sites deemed most important to business users. These measurements are a natural complement to the data Web-Sniffer produces on its own and provide another way of assessing if problems are based in the network or on the Web server host.

The four functions of the WebSniffer repository are:

- Collecting, correlating, and analyzing WebSniffer agent data
- Generating alarms based on problems it identifies during data analysis
- Storing collected WebSniffer agent data and configuration data
- Acting as a Web server for the Java-based user interface

To identify problems with Web sites, the repository applies internal rules to the data the WebSniffer agents collect. Once a problem is identified, the repository includes the problem on a list of active alarms so that a user can see what has occurred. The WebSniffer repository can also generate external alarm notifications such as sending a message via e-mail, sending a message to a pager, or sending an SNMP trap. Additionally, a scripting capability allows users to specify their own external alarm notification actions. All alarming capabilities are accessible from the information center. Alarm conditions, product configuration, and data retrieved from the agents is stored in the repository for viewing by the user interface. Data are accessible for 24 hours. However, alarms are stored longer so that even when data can no longer be viewed, a record of the problem exists. Because the user interface is browser-based, the repository also acts as the Web server that serves the user interface.

The final component is the information center. It represents the graphical user interface (GUI) written in Java that will run in a browser. The GUI uses a tab metaphor for quick switching between functional areas, and to allow data to be grouped and sorted hierarchically. Its three main areas of functionality are:

- Alarm screen
- Data drill-down areas
- Expert solutions guide

Network Associates offers a combination of Web site analysis tools that go far beyond just monitoring communication links.

6.4.2 TELEMATE.NET FROM TELEMATE SOFTWARE

Firewall managers obtain plenty of data regarding Internet user access — hundreds of megabytes of firewall traffic logs each month. Although the log is an important resource, the large volume of cryptic and redundant data makes it difficult to readily extract the information. In its raw form, a firewall log may consist of millions of lines detailing each unique element that passed through the firewall. This may generate several pages of data for even a single visit to a Web page. When multiplied by hundreds of users accessing many Web sites, volumes become unmanageable. Even the physical disk space required to store log files becomes an obstacle. Also, the log file lacks the information to relate abstract traffic records to a department or company. With these constraints, it is nearly impossible for a firewall manager to draw meaningful perspectives out of the log. Early attempts to make use of log files revolve around PERL scripts or homegrown databases. These approaches suffer from many setbacks, including high setup and ongoing maintenance requirements and performance concerns. In addition, the reporting interfaces usually aren't detailed enough to meet users' requirements.

Network management platforms and systems management applications provide a technical view of the traffic flows. These systems are primarily designed to measure network and systems elements such as segment utilization, packet routing, fault management, and response times. They do not, however, provide insights to the actions of individual users or help the firewall manager assess the business value of the traffic generated.

Telemate.Net translates abstract data output from firewalls and other network sources into business reports that serve as the basis for effective decisions. Its data collection and compression capabilities allow managers to collect and store several months' worth of Internet connection information for accurate trend analysis. By funneling abstract Internet traffic records to a corporate directory of users, departments, and companies, Telemate.Net creates a valuable basis to understand and manage Internet use.

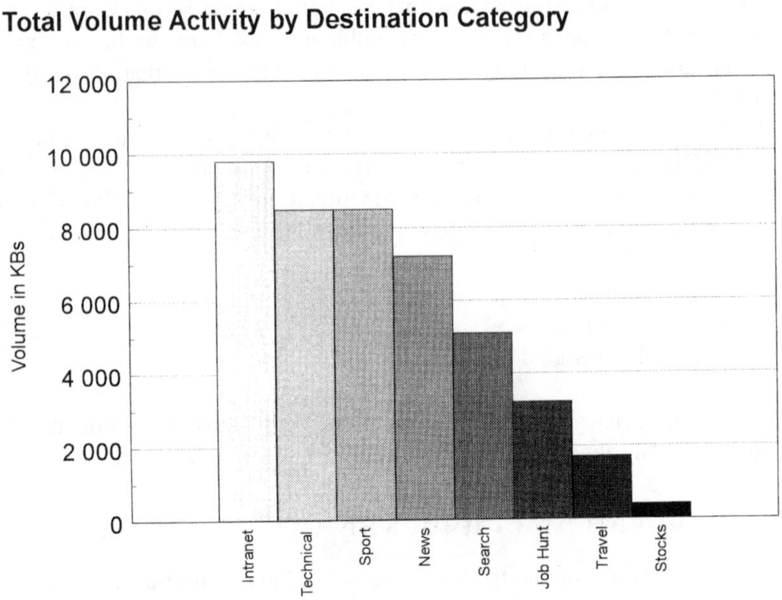

Total Volume Activity by Destination Category

FIGURE 6.8 Total volume activity by destination category.

6.4.2.1 Functions of the Product

Telemate.Net resides on Windows95 or NT machines and reads firewall logs. It supports input from multiple firewalls and proxy servers, even if they are from different vendors. This multisource approach makes it easy for Telemate.Net to adapt to real-world customer requirements. Once the data is received, Telemate.Net automatically processes the raw data, calculates usage costs, and associates the traffic records with a corporate directory of users, departments, and companies. Customization is supported for costing by kilobytes transferred, number of files, protocol, time of day, and day of the week. Also, fixed costs may be assigned to users. This flexible approach helps firewall managers replicate the actual ISP billing method or incent system users to avoid needless Internet activity.

It also applies complex algorithms to relate hundreds of traffic records into more manageable views. It offers users the ability to view information summarized by Web page accesses. This helps to overcome the data overload presented by the firewall log. It also makes it practical to store and report on several months' worth of activity, providing a more valid method of assessing traffic patterns and assigning costs. Automated reporting tools transform the large amount of data into concise management reports. These reports may be distributed to users via print, e-mail, or HTML on a recurring schedule. There are more than 40 standard reports, each of them containing multiple filters for users, sites, type of traffic, etc. An integrated custom report writer also provides extended abilities to modify reports or create new ones. Ad hoc reporting is supported as well. Figure 6.8 shows the total volume activity by destination categories.

TABLE 6.5
Ten Most Active Sites by Volume

No.	Site Name	Total Volume (KB)	% of Overall Volume	Number of Accesses	Cost
1	Site 1	502.6	13.2	12	$0.47
2	Site 2	450.4	11.8	9	$0.38
3	Site 3	364.9	9.5	13	$0.28
4	Site 4	308.8	8.1	8	$0.29
5	Site 5	272.1	7.1	4	$0.10
6	Site 6	264.7	6.9	2	$0.26
7	Site 7	226.5	5.9	6	$0.13
8	Site 8	225.2	5.8	6	$0.14
9	Site 9	198.5	5.2	2	$0.19
10	Site 10	125.8	3.3	4	$0.12

TABLE 6.6
Ten Largest Downloads by Volume

No.	Total Volume (KB)	Destination	Date	Time	Cost	Protocol	Category
1	280	www1.com	5/6/99	08:11 AM	$0.84	HTTP	Sports
2	265	www2.com	5/6/99	03:13 PM	$0.79	HTTP	Technical
3	245	www3.com	5/6/99	02:47 PM	$0.67	HTTP	Travel
4	195	www4.com	5/6/99	10:23 AM	$0.51	HTTP	News
5	145	www5.com	5/6/99	10:55 AM	$0.43	HTTP	News
6	135	www6.com	5/6/99	01:11 PM	$0.39	HTTP	News
7	135	www7.com	5/6/99	07:47 PM	$0.39	HTTP	Sports
8	100	www8.com	5/6/99	11:15 AM	$0.27	HTTP	News
9	95	www9.com	5/6/99	10:00 AM	$0.25	HTTP	Technical
10	90	www10.com	5/6/99	08:00 PM	$0.22	HTTP	News

Raw data can be used to quantify site activity, downloads, and top accessed sites. In all cases, the total volume is the basis of quantification. Table 6.5 shows the most active sites. The largest downloads are displayed in Table 6.6. Finally, Table 6.7 displays top accessed sites by total volume.

6.4.2.2 Advantages of Using the Product

Several components must come together to build an effective source of business information. Without this implementation, a usage management system could fail to meet customers' needs. The Telemate.Net approach has several practical advantages:

TABLE 6.7
Top Accessed Sites by Total Volume

No.	Site Name	Number of Accesses	Unique Users	Volume In (KB)	Volume Out (KB)	Total Volume (KB)	% of Overall Volume
1	Site 1	146	12	9215.6	378.0	9393.6	16.4
2	Site 2	143	12	6882.0	161.0	7043.0	12.1
3	Site 3	104	10	5181.3	146.8	5328.1	9.1
4	Site 4	78	13	3358.4	76.7	3445.1	5.9
5	Site 5	61	10	2915.3	68.1	2983.4	5.1

- Use of multiple data sources — It can accept data from a variety of firewalls and proxy servers, even those from different vendors. This allows Telemate to adapt to real-world requirements such as multi-site management, support for legacy systems, and complex networks. By integrating data input and information management into a single interface, this product saves money, reduces training requirements, and provides more complete reporting.
- Scalability — An effective information system is useless if it collapses under high volumes of data. Telemate.Net is able to translate massive amounts of data into concise business reports. This allows users to see views of site accesses and permits storage and reporting on several months' worth of data. This provides more relevant information for analysis and decisions.
- Use of cost allocation — Telemate.Net calculates costs on Internet activity and allows companies to allocate these costs to serve organizational purposes. It also serves as an effective tool to help control needless use.
- Organizational directory — Telemate.Net takes this database of abstract Internet traffic records and links them to a directory of real people, departments, and companies. This provides an easy way to see the sites hit and who hit them and prepares the information for management use.
- Flexible reporting — Any system can report on all the traffic. The true distinction is the ability to easily extract the focused facts the users need and protect receivers of reports from overload. Telemate.Net offers a number of standard reports that allow users to quickly produce professional results. In addition, flexible, multilevel filtering and custom report writing tools help users get concise, focused information adapted to their specific needs.
- Support of automated operations — Telemate.Net automatically collects data and generates and distributes reports. It also manages routine system housekeeping tasks. By putting the system on auto-pilot, the firewall manager saves time of repetitive work. It also ensures the reliable flow of activity reports to the right addresses.

- Simple hardware and software requirements — Users do not need to reconfigure networks and purchase specific hardware. Its operation is completely independent of critical Internet traffic flows. It will never cause a performance bottleneck on the user's network. The product can be operational within a few minutes. Its interfaces enable a broad use of the product without a lot of training and specialized knowledge.
- Increase of Internet security — Internet access and use must be protected. Telemate.Net records and monitors excessive and unauthorized data transfers and highlights abnormal patterns. Reports can identify unusually large FTP transfers or e-mail attachments to alert managers to suspicious outgoing information, such as unauthorized transfer of software source code.

Telemate.Net is unique in the sense that it uses firewall input data to report on Internet and intranet resource usage patterns.

6.4.3 NET.MEDIC PRODUCT FAMILY FROM VITAL SIGNS

Net.Medic works with a browser to monitor, isolate, diagnose, and correct Internet or intranet performance problems. It identifies problems rapidly, offers recommendations for solving them, and in many cases, automatically fixes them. It is also a powerful desktop agent, which, when used with other server-based products, gathers, analyzes, and reports on the user's Internet/intranet experience, helping system providers respond quickly to performance problems and design more effective networks.

It animates the end-to-end connection, showing the user exactly what is occurring across the Internet or intranets, including traffic jams and bottlenecks. Color indicators highlight the responsiveness of each component. And as the users move from page to page, Net.Medic arrives split seconds earlier, continuously monitoring the unique path. Once the problem is identified, Net.Medic provides the administrator with critical vital signs about the health of the affected component.

By continuously monitoring activity along the end-to-end connection, Net.Medic is able to tell exactly what is happening at every critical juncture. It reports throughput, retrieval time, the Web server load and efficiency, as well as network delays and congestion levels. Net.Medic also monitors the PC and modem, ensuring optimal configuration for real-time Internet pathway conditions and providing historical trend information that enables the administrator to make educated decisions about the Internet/intranet use.

Net.Medic automatically eliminates performance problems before users encounter them. This automation may include reinitiating server requests and modem compression that eliminates long wait times and time-outs. This feature is supported by AutoCure. Figure 6.9 shows a status overview with a number of different metrics, such as end-to-end overview, data transfer rates, Web page retrieval time, PC performance, modem compression, traffic levels, and delays.

In case of threshold violations, Net.Medic can automatically generate an e-mail message to the appropriate contact requesting assistance. It also monitors and reports the performance of ISPs, logging busy signals, call completions, disconnects, average connect rates, and other data about the service level users actually experience. It

Net. Medic Performance Metrics

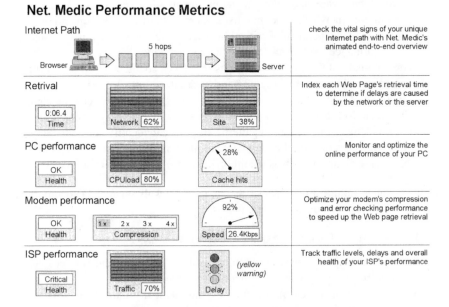

FIGURE 6.9   Net.Medic status screen.

continuously monitors the ISP connection and Internet conditions, dynamically adjusting the PC and modem configuration to optimize Internet performance. Figure 6.10 provides history reports.

Net.Medic Pro gives all the functionality offered by Net.Medic. However, whereas Net.Medic traces individual users, Net.Medic Pro lets administrators continuously monitor the ongoing performance of each Internet/intranet component as well as how they work together. The result is a powerful diagnostic tool.

The automonitor feature automatically monitors application performance over LANs, WANs, and dial-up connections, allowing users to view network transactions as they occur. For example, users can automatically monitor the availability and response time of Web servers by typing in the URLs and defining the testing intervals. Automonitor automatically accesses the servers according to the schedule users define. The health log then summarizes the data about any problems encountered during testing, along with the regular, ongoing data collected by Net.Medic Pro. In addition, the automonitor can be used to validate service level agreements with end users. The notify feature alerts administrators when a performance-affecting problem occurs during an automated test sequence. An e-mail message is immediately generated, enabling the administrator to proactively respond and minimize the impact on users. The status log provides underlying details of the problem and its causes, giving users the necessary information to take corrective action. It also contains a detailed record of all testing transactions, enabling the operator to review the details and progress of automonitor tests. Its history reports highlight chronic bottlenecks, peak usage periods, and network trends, offering users the information basis to maintain high quality of service.

Net. Medic Performance Reports

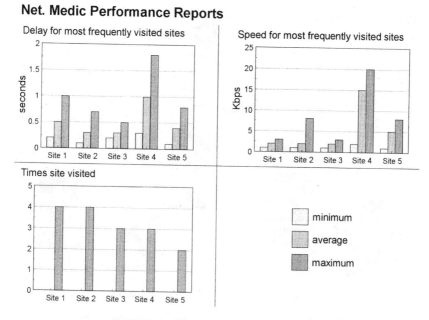

FIGURE 6.10 Net.Medic history reports.

The structure of Net.Medic Pro is shown in Figure 6.11. By installing the product at multiple sites on a corporate intranet, IT professionals can automatically monitor Web site performance from different access points. ISPs can use Net.Medic Pro to regularly test their customers' Web sites and to ensure high quality of service. Data from multiple access points enables the administrator to compare application performance as experienced by customers accessing applications from different areas of the intranet, track Web server availability and response, identify leading indicators of emerging issues, and proactively address problems. Detailed diagnostic data is available from the Net.Medic Pro health log if any anomalies are detected during the test sequence.

The automonitor feature can also be used to regularly dial into modem banks to test the availability and performance. Detailed data, such as log-in times, time to connect, connection speeds, busy signals, and other call failures, are provided to track service levels, ingress and egress speeds, and network throughput from the user's perspective. ISPs and IT professionals can combine Net.Medic Pro's automatic Web site monitoring and dial-up diagnostic capabilities to understand end-to-end performance in various geographic areas. Figure 6.12 shows a realistic combination with ISPs and the Internet.

By dialing into modem pools at different points of presence (POPs), ISPs can access their own critical Web servers, as well as users' most popular Web sites. Net.Medic Pro data provides valuable information about network availability and throughput for users accessing various segments of the network.

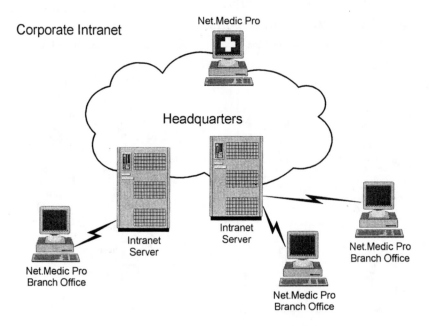

FIGURE 6.11   Distributed monitoring with Net.Medic Pro.

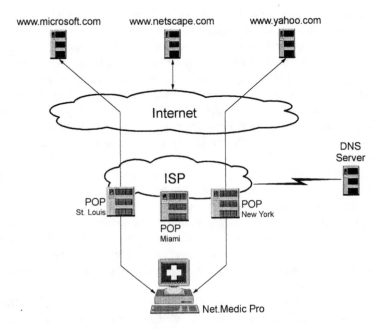

FIGURE 6.12   Monitoring and testing connections with Net.Medic Pro.

Visual OnRamp Reports

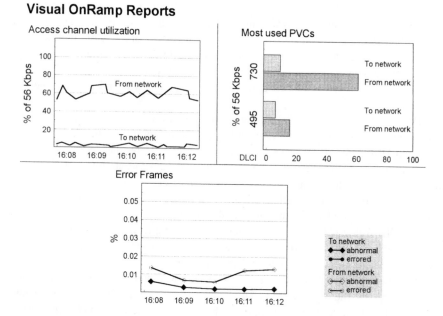

FIGURE 6.13   Visual Onramp reporting examples.

6.4.4   ONRAMP FROM VISUAL NETWORKS

Tracking Internet use is a high-priority item for network managers. Visual Networks is helping to simplify this problem with Onramp CSU/DSU, which features an integrated IP protocol analyzer. It tracks network utilization at all layers of the OSI stack. Statistics are accessed from remote PCs running management software provided with the product. Because Onramp performs full real-time decodes, it can be used to troubleshoot IP performance problems from the physical layer up to the application layer. Since Onramp stores collected statistics in an SQL format, network managers can easily create extended network baselines that can help predict bandwidth shortages before they occur. Onramp is SNMP based, so third-party network management applications can use it for basic tasks such as alarm monitoring.

Onramp consists of a CSU/DSU and requisite application management software that is installed on PCs running Windows95 or NT. The CSU/DSU integral protocol analyzer collects operating statistics every 15 minutes and stores them. Management statistics are downloaded daily through a direct Ethernet or Token Ring connection or out of band via a modem connection. Daily downloads average about 50 KBs. The integrated IP analyzer can collect many statistics, but the emphasis is on Internet utilization. It simultaneously displays top talkers, Internet utilization as percentage of WAN bandwidth, Internet throughput by protocol, and overall protocol distribution. Figure 6.13 shows a couple of examples (Taylor, 1996).

Top talkers list Internet usage percentages by IP or Web address for machines sending information to and hosts transmitting information from the Internet. Similarly, when Internet utilization is tracked as a percentage of available bandwidth, it is done for outgoing and incoming traffic. Visual Onramp includes an automatic report generator that creates bill-back charts and baselines.

The product consists of two components: the DSU/ASE and the management application.

DSU/ASE — This is the managed Internet access device. It combines the functionality of a protocol analyzer and transmission monitor into a fully SNMP-managed DSU/CSU. It replaces a traditional DSU/CSU on the dedicated Internet connection, interfacing between the router/firewall and the circuit to the Internet. Models are available for either 56 Kbps or T1/FT1 access. It employs a dedicated high-speed processor that collects and stores network operational data about all Internet traffic. Its CSU/DSU function is fully compliant with standards. Its extensive SNMP MIB follows RMON2 with extensions specific to Internet monitoring. Full seven-layer analysis is performed on every Internet packet, from the physical and frame relay layers to the Internet application layer.

The management application — This is deployed on one or more management workstations on the network. The application retrieves specific operational data collected by the DSU/ASE and intelligently analyzes it to provide meaningful management information up through the protocol stack. An integral database archives Internet operational data to facilitate long-term report generation and historical performance analysis. The application is organized around three tool sets:

1. Usage accounting — It includes application profiling tools that rapidly show Internet usage patterns to determine how access bandwidth is being used. Actual usage may be compared with targeted usage. Detailed usage information allows planners to evaluate activity, needs, and trends, as well as to measure progress toward business goals. Many organizations need a method of properly allocating network costs back to various departments or workgroups for their Internet usage. Onramp provides the data collection and report generation for accurate billing.

2. Problem resolution — The Internet is becoming a mission-critical business tool, so it is imperative that connectivity is continuous and of the highest quality. Onramp provides a full range of problem resolution tools to ensure that Internet access is always there when users need it. Proactive monitoring helps to avoid many problems altogether. Proactive monitoring concentrates on network irregularities and on what might be causing them. Rapid troubleshooting helps to quickly correct poor network performance before the degradation gets too serious.

3. Bandwidth planning — Predictions of Internet usage are very difficult. The best way is to follow usage trends and to plan for Internet access bandwidth accordingly. Accurate and timely information is required to make these strategic bandwidth decisions. Onramp helps with simple planning reports. Detailed historical information about Internet access performance is collected continuously and archived in a relational database.

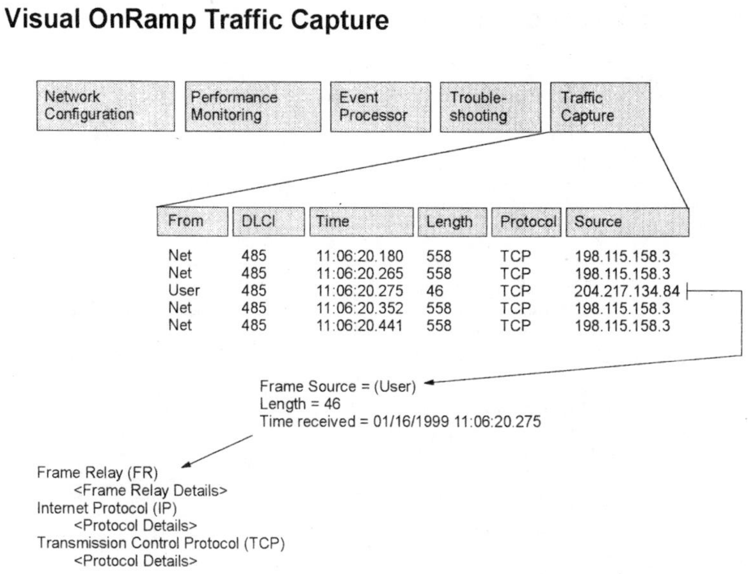
Visual OnRamp Traffic Capture
Network Configuration
Performance Monitoring
Event Processor
Trouble-shooting
Traffic Capture
From
DLCI
Time
Length
Protocol
Source
Net 485 11:06:20.180 558 TCP 198.115.158.3
Net 485 11:06:20.265 558 TCP 198.115.158.3
User 485 11:06:20.275 46 TCP 204.217.134.84
Net 485 11:06:20.352 558 TCP 198.115.158.3
Net 485 11:06:20.441 558 TCP 198.115.158.3
Frame Source = (User)
Length = 46
Time received = 01/16/1999 11:06:20.275
Frame Relay (FR)
<Frame Relay Details>
Internet Protocol (IP)
<Protocol Details>
Transmission Control Protocol (TCP)
<Protocol Details>

FIGURE 6.14 Frame relay summary by Onramp.

Figure 6.14 displays traffic evaluation results for frame relay, IP, and TCP. Troubleshooting can also be supported. Figure 6.15 shows V.35 signal activity and DDS errors.

Due to its unique data capturing interface, Onramp is an interesting product for traffic analysis and network planning. Chapter 9 offers more solutions and products.

6.4.5 IntelliFlow from Resonate

The IntelliFlow architecture introduces an intelligent solution for enterprise traffic management, delivering a flexible, scalable system for optimizing, coordinating, and managing TCP traffic across multi-tier networks. The IntelliFlow architecture provides scheduling capabilities at each level of the infrastructure: resonate global dispatch software for the WAN level, resonate central dispatch software for the Web level, and resonate application dispatch software for application servers. Enterprise-wide monitoring and centralized management software integrates dispatch products to provide event identification and further intelligence for scheduling traffic across the enterprise. This product suite enables companies to realize the power, flexibility, and full potential of Web-centric computing.

The IntelliFlow architecture provides a complete and powerful solution for enterprise traffic management of multi-tier networks. It is an integrated software solution for all levels of the network infrastructure. This is a suite of products that work together to collect information about system status and direct TCP traffic to the servers best suited to respond to users' requests. It transforms a distributed, multi-tier corporate infrastructure into a cohesive system that intelligently and dynamically directs traffic to the optimal resource or server to deliver the fastest and

Visual Troubleshooting

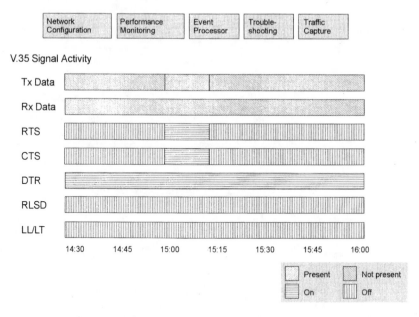

FIGURE 6.15   Troubleshooting with Onramp.

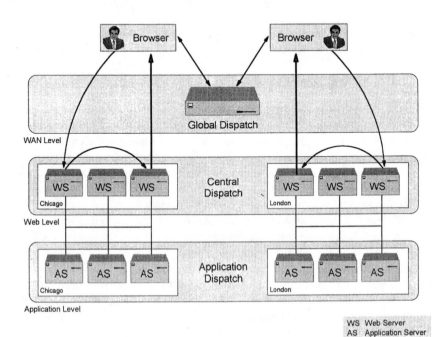

FIGURE 6.16   IntelliFlow architecture.

most reliable access. Figure 6.16 shows the IntelliNet architecture. The product consists of five components. They will be described as follows.

6.4.5.1   Resonate Global Dispatch

This is an authoritative DNS solution that resides on network DNS servers and provides intelligent, WAN-based scheduling capabilities. Featuring real-time monitoring and intelligent traffic-direction policies not found in other products, the resonate global dispatch directs each IP request to the POP offering the best performance for the end user initiating the request. It is the only WAN scheduling product that considers the three most important metrics when deciding where to schedule a request:

- Client/server latencies
- Real-time server load
- Server availability

Global dispatch agents at each POP monitor the POP's load and availability and measure the response time from the users' local DNS server to the POP. A global dispatch scheduler on the DNS server collects the latency measurements and status information from the agents, and then returns the IP address (or virtual IP address) of the logically closest, least loaded, available POP.

Resonate global dispatch enables network administrators and ISPs to deliver high availability and high performance to users on distributed networks. Network administrators can use global dispatch to establish traffic policies and priorities; for example, directing requests from all European field offices to a European POP. Figure 6.17 shows global dispatch in operation.

6.4.5.2   Resonate Central Dispatch

Resonate central dispatch provides intelligent Web-level scheduling of TCP-based services. Central dispatch enables a collection of servers to appear as a single, reliable system accessed with a virtual host name. Using a patent-pending resource-based scheduling technology, central dispatch directs each TCP request to the server best suited to respond to that particular request. A single central dispatch site can manage service requests for HTTP, FTP, SMTP, and other TCP services.

Using resonate central dispatch, an IT organization can deploy multiserver Web sites, each of which appears to users as a single virtual site, even though its servers may actually be distributed across subnets and firewalls. The servers at a virtual site need not be mirrored; each can store its own data. It provides immunity to server failures by detecting them and directing traffic around failed systems. Unlike simple load-balancing solutions, the central dispatch software-only solution eliminates single points of failure and does not require dedicated hardware.

Resonate central dispatch ensures that end users receive the fastest possible response to requests. Web sites achieve near-linear performance increases as Web servers are added to a central dispatch site. The structure is shown in Figure 6.18.

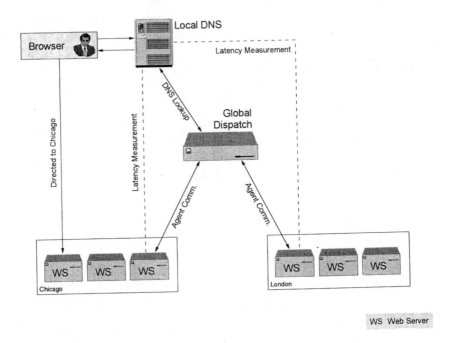

FIGURE 6.17 Global dispatch in operation.

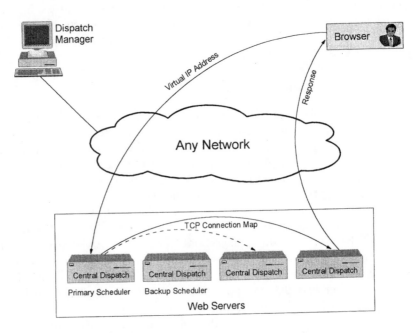

FIGURE 6.18 Central dispatch in operation.

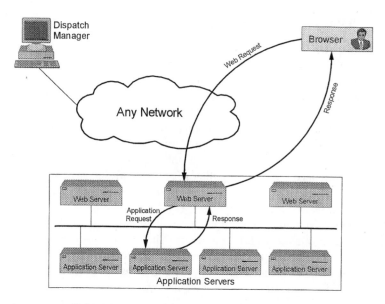

FIGURE 6.19   Application dispatch in operation.

6.4.5.3   Resonate Application Dispatch

Resonate application dispatch provides intelligent scheduling for CGI and Web server API calls to back-end application servers. It enables farms of application servers to respond to requests from farms of Web servers. Application dispatch works with central dispatch to direct traffic intelligently, ensuring that Web applications always receive the back-end support they need. Should an application server, such as a database server, fail, application dispatch directs application requests to another server that is able to respond. This dynamic traffic redirection is transparent to users. Users experience uninterrupted service from applications, without having to monitor or select application servers themselves. Application dispatch ensures that application services are always available, and that high-priority service requests receive the highest attention. Application dispatch supports industry-standard Web application interfaces, including CGI, NSAPI, ISAPI, and Apache modules. Figure 6.19 shows application dispatch in operation.

6.4.5.4   Resonate Dispatch Manager

Resonate dispatch manager provides integrated Java-based management of the entire IntelliFlow system. Using the dispatch manager's GUI, system administrators can configure virtual Web sites, allocate software resources, and establish traffic policies for the network. The dispatch manager can reside anywhere on the network and runs on any Java-ready platform.

6.4.5.5   Resonate Dispatch Monitor

Resonate dispatch monitor provides monitoring and control of all enterprise resources. Dispatch monitor works with the dispatch manager to respond to real-time events. Information collected by the dispatch monitor is fed back into the IntelliFlow architecture, enabling dispatch servers to direct traffic around system failures and overloaded POPs. The result is a dynamic traffic redirection system that provides business operations with continuous access to the resources they need.

When events occur, such as server failure, the dispatch monitor captures the event and notifies the dispatch manager. It then takes appropriate action based on specified, user-defined scheduling policies. For example, if an application server fails at a POP, global dispatch can be notified to send requests to another available POP which can service them.

The IntelliFlow architecture simplifies the management of enterprise networks. It provides administrators with real-time data about system performance and provides a single point of control for implementing scheduling policies at each point in the system. Dispatch manager provides a straightforward, graphical user interface for managing distributed, multi-tier infrastructures.

6.4.6   NETSCOPE FROM APOGEE NETWORKS, INC.

NetScope combines strong real-time monitoring capabilities with state-of-the-art reporting features. It is a remarkably logical and innovative network management tool that has proven to be effective in some of the world's most challenging corporate environments. By giving networking staff instant access to a complete range of traffic and performance data via highly intuitive graphical presentations, NetScope makes it easy for even relatively inexperienced team members to quickly pinpoint problems and anomalies. It also provides flexible caching and historical data for critical after-the-fact analysis.

NetScope learns all about the protocols and applications on networks. So, it doesn't require extensive configuration before it starts delivering value; just as important, it lets users customize access privileges to match each team member's job function and needs. Users will eliminate counterproductive screen clutter and be able to put the right real-time network data in front of the right people — whether they are in operations, performance analysis, systems engineering, or design and development.

Its architecture fulfills all the essential requirements of an enterprise-class network monitoring solution, including:

- Fully distributed, client/server design for maximum scalability
- Browser/Java client for anywhere, anytime management
- Rich data access controls for superior security
- Comprehensive reporting and trending tools with rich drill-down capabilities
- Full integration with existing industry-standard data sources, such as RMON1, RMON2, SMON, and NetFlow
- Totally intuitive, context-sensitive point-and-click operation

A typical configuration consists of a NetScope server, a report server, databases, and a number of NetScope clients. NetScope servers are connected to managed objects that usually support RMON1, RMON2, SMON, and NetFlow. NetScope servers poll these data sources, interpret data, and forward results in formatted records to the report server. Both types of servers are accessible by universal browsers.

Besides meeting these requirements of enterprise-class monitoring, NetScope may be differentiated from other wire monitors by the following features:

- Visual representation of network traffic data: Users are able to instantly determine "top talkers" by segment, protocol, or application. Anomalies, spikes, and trends in host activity will be immediately discernible. Graphical presentation of data, combined with intuitive, context-sensitive navigation, puts all relevant information right on the operating console.
- Elastic real-time cache: NetScope automatically retains status information during periods of anomalous network performance. Users can determine the conditions of the other applications or hosts or other network vital signs when the problem occurred. This powerful capability lets users see exactly what was happening on the network during performance problems, even if that problem no longer exists, even if it occurred several hours ago.
- Automatic discovery and self-configuration: With NetScope, users don't need either the time or the experience to define filters and probes. In order to set filters, users should know exactly what is flowing through the network. The filtering process is automated and self learned. NetScope discovers all protocols and popular applications running on the network, and it sets up every screen to display the relevant statistics. If a new protocol shows up on the network, NetScope lets users know right away. So NetScope starts delivering value immediately without extensive learning curves or configuration hassles.
- Flexible, customized views for different users: Much of the product's value and power come from its ability to present users with the specific data they need to do their jobs. The product helps engineers assigned to support individual business units to focus on their area of responsibility and nothing else. The same is true with network architects. They receive views about protocol distribution and bandwidth utilization trends. NetScope can partition the views to satisfy operating, administration, and design and development, at the same time.

NetScope empowers users to see exactly how intranets are performing — both in real time and historically — rather than merely monitoring a collection of hardware devices. As a result of this "performance-centric" network view, bottom-line benefits are:

- Faster, more precise root-cause analysis of both chronic and intermittent network problems
- Rapid empowerment of operations staff at all skill levels

- Significantly improved capacity planning and problem prevention
- Reduced cost of network operations

In summary, NetScope is a foundation product to help users deliver service level management, application performance, and switch management functionality with intranets.

6.5 SUMMARY

Traffic measurements are very helpful to provide performance metrics. The measurement technology differs from log file analyzers, but many metrics are the same or similar. It is expected that multiple and different kinds of tools will be connected together. Traffic measurement tools may be combined with response time measurement tools offering end-to-end management. Resonate and Freshtech are working together to offer this type of an integration. This integration step may be followed by integration with management platforms.

7 Web Server and Browser Management

CONTENTS

7.1 INTRODUCTION

The content of Web pages is maintained on Web servers. Usually, these are processors running under Unix or NT. They must be flexible and scalable enough to cope with significant workload fluctuations.

Server management is comprised of several functions (Sturm, 1998):

- Server monitoring — This is the basic component of server management. It requires someone or something to keep a constant watch on the status of the managed servers. Fortunately, much of this painfully tedious task for human beings can be automated by management platforms, such as Unicenter TNG or HP OpenView. Monitoring is essential for detecting problems as soon as they occur, and for gathering data for use in performance management.
- Workload management — This function consists of scheduling and tracking the jobs that run across one or more servers in a heterogeneous environment. Workload management takes into account calendar requirements such as time of day, day of week, or holidays. It also considers dependencies between workloads, such as "Job A must be finished before Job B can be started," as well as what to do in the case of a failure.
- Server performance management — Its purpose is to ensure that servers are working efficiently. The keys to this function are data collection and trend analysis.
- Server capacity planning — Whereas the performance management function focuses on current effectiveness, capacity planning ensures that servers will work effectively in the future. The keys to this function are historical analysis and forecasting.

7.2   CRITICAL ISSUES WITH WEB SERVER MANAGEMENT

The management architecture for Web servers may be centralized or decentralized, or a combination of both. A centralized solution (Figure 7.1) assumes that all Web servers can be managed from one location. When the number of Web servers to be managed exceeds a certain number, this solution could become critical in terms of networking overhead. It is assumed that with the exception of collecting raw data, all processing functions are executed in the manager.

With a decentralized solution, domain managers take over the responsibility of managing a certain number of Web servers (Figure 7.2). Each domain is actually a centralized solution on its own. Domain managers may communicate with each other or can even be connected to an umbrella manager. Network overhead can be well controlled and kept to a minimum. Domain managers usually exchange only consolidated data with each other. The result is that the communication overhead can be kept to a minimum.

Practical arrangements usually work with a combination of these two alternatives. If umbrella management is the choice, the manager can also manage other components, such as switches and routers, and can correlate data with server management.

Web servers are usually deployed on Unix or NT platforms. It is important to know whether or not these two platforms can be cross-managed.

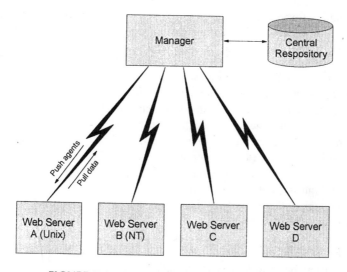

FIGURE 7.1   Centralized Web server management.

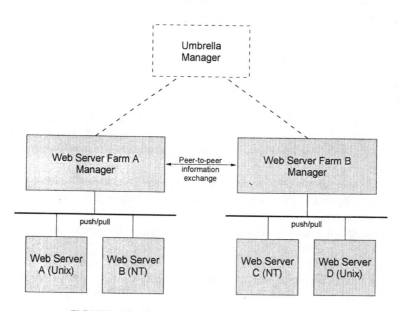

FIGURE 7.2   Decentralized Web server management.

In terms of management hardware and software, there are multiple choices. The software is UNIX or NT; both are working with a number of hardware platforms.

Data capturing techniques are critical for both overhead and performance of the management architecture. Measurement probes or agents are located inside the operating system; they run with relatively high priority and can supervise both hardware and software components of Web servers. Raw data are expected to be stored away immediately. Processing can be done in the Web server or in the manager. The targeted metrics to be collected include:

- What is the CPU utilization by applications?
- What are the physical and logical I/O rates?
- Can the list of active applications be generated?
- What is the average queue length for the CPU?
- What is the average queue length for I/O devices?
- How high is the CPU/I/O overlap?
- Are process wait times measured and displayed?
- How high is the disk utilization?
- How high is the memory utilization?
- Are swap rates measured?
- What resources are processes blocked on?
- What reporting is used?
- Can users be identified by application?

Raw data or preprocessed data are stored at the Web servers with the intention of being uploaded for further processing by the manager. Upload may be controlled in one of two ways:

- By events such as filling a percentage of storage space, or time, or when critical data are captured
- By polling cycles initiated by the manager

Both alternatives have pros and cons; the selection depends on the actual configuration, data volumes, and communication protocols in use. Web server management can use SNMP for transmitting data, assuming Web server metrics are stored and maintained in MIBs. Another alternative is the use of DMI-like standards for storage and transmission. The most recent alternative is the use of embedded Wbem agents that support CIM for storing and exchanging data. In this case, HTTP is the protocol of choice.

Overhead is a critical issue with large data volumes. Data capturing is expected to introduce little overhead, when data are stored immediately. If local processing is taking place, overhead must be very carefully quantified; if resource demand is high, overall Web server performance may be impacted. Data transmission overhead can be heavy if everything is transmitted to the site where processing is taking place. WAN bandwidth is still too expensive to be dedicated just to transmitting measurement data. If bandwidth is shared with other applications, priorities must be set higher for business applications than for transmitting raw log file data.

Data losses cannot be completely avoided. Data capturing functions in Web servers, storage devices, or components of the transmission may fail; in such cases, there will be gaps in the sequence of events. Back-up capabilities may be investigated, but IT budgets won't usually allow too much spending for backing up large volumes of performance data. In the worst case, certain time windows will be missed in reporting and statistics. Those gaps may be filled with extrapolated data.

Due to considerable data volumes, databases should be under consideration to maintain raw and/or processed Web server measurement data. Database managers would then offer a number of built-in features to maintain measurement data. Clustering visitors may be deployed from various perspectives, such as geography, common applications, common interests on home pages, data, and time of visits. Automatic log cycling can also be supported here by the database managers. ODBC support helps to exchange data between different databases and to correlate data from various databases. Besides measurement data, other data sources can also be maintained in the same data warehouse. Data analysis with concrete targeted reports as well as special analyses may occasionally be conducted. This type of analysis, called data mining, can discover traffic patterns and user/visitor behavior. Both are important to sizing systems and networking resources.

One of the most important questions is how Web server management performs when the number of managed Web servers — and as a result of this, data volumes — increase. All resources, such as processors, storage devices, I/O devices within the Web servers, and networking components, may become pary of the bottleneck. In any case, collection and processing capabilities must be estimated prior to deciding on procedures and products.

To reduce processing and transmission load of measurement data, redundant data should be filtered as near as possible to the data capturing locations. Filters can help avoid storage of redundant data. Filters can also be very useful in the report generation process. Again, unnecessary data must not be processed for reports. Powerful filters help to streamline reporting.

Not everything can be automated with measurement data analysis. The user interface is still one of the most important selection criteria for products. Graphical user interfaces are most common, but simple products are still working with textual interfaces. When Web server management is integrated with management platforms, this request is automatically met by management platforms.

Reporting is the tool to distribute the results of Web server management data analysis. Predefined reports and report elements as well as templates help to speed up the report design and generation process. Periodic reports can be automatically generated and distributed for both single Web servers and Web server farms. In the case of many Web servers, report generation must be carefully synchronized and scheduled. Flexible formatting helps to customize reports to special user needs.

There are many output alternatives for reports. The most frequently used solutions include Word, Excel, HTML, and ASCII. Report distribution also offers multiple choices:

- Reports may be stored on Web servers to be accessed by authorized users who are equipped with universal browsers.
- Reports can be uploaded into special servers or even pushed to selected users.
- Reports may be distributed as attachments to e-mail messages.
- Reports can be generated at remote sites; this alternative may save bandwidth when preprocessed data instead of completely formatted reports are sent to certain remote locations.

TABLE 7.1
Selection Criteria for Web Server Management Tools

Architecture of the product
  Centralized
  Decentralized
  Combination
What servers can be managed
  Unix
  NT
  Others
System requirements
  Hardware
  Operating system
Data capturing techniques
  Locations of probes
  What metrics can be collected
  Storing collected data
  Upload to processing (eventing or polling)
Overhead
  Data capturing
  Data processing
  Data transmission
Precautions against data losses
Database management
  Use of a database manager
  Grouping capabilities
  ODBC support
  Data warehouse
  Drill-down (mining) capabilities
Scalability
Flexibility
Filtering
  Reports
  Information within reports
User interface
  Textual
  Graphical
  Integration with management platforms
Reporting
  Predefined reports
  Use of templates
  Predefined report elements
  Automated server farm reporting
  Trend reporting
  Use of a report scheduler
  E-mail reports
  Report output alternatives (Word, Excel, HTML, ASCII)

TABLE 7.1 (continued)
Selection Criteria for Web Server Management Tools

Reporting (continued)
    Flexible formatting
    Cross-tabs by correlating fields in the database
    Customization capabilities
Distribution of reports
    Accessing reports
    Uploading reports
    Remote reporting
Integration with management platforms
Documentation and support
    Integrated online manual
    Paper manual
    Help desk
Vendor
    Number of clients
    Founded when
    What other products are offered
    Support (e.g., on-site, hotline)
    Maintenance contracts
    Financial strength
    Keeps current with servers software releases

Documentation may take various forms. For immediate answers, an integrated online manual would be very helpful. Paper-based manuals are still useful for detailed answers and analysis. This role, however, will soon be taken over by Web-based documentation systems. In critical cases, a hotline can help with operational problems.

Managing Unix and NT servers represents just another management application. If management platforms or umbrella managers are used, these applications can be integrated into the platform. There are many ways to integrate; the most common is to deploy a command line interface (CLI) solution. Integration may even be supported by a management intranet. Every participant is equipped with a universal browser and communicates with management applications that reside in managed objects and that are equipped with lean Web servers. Table 7.1 summarizes selection criteria for Web server management tools.

The majority of Web server implementation is based on Unix or NT. Some are on NetWare, but their market share is not significant.

7.3 MONITORING UNIX WEB SERVERS

There are a growing number of Unix systems management applications now being developed for or ported to SNMP-based management platforms. This indicates the convergence of network and systems management. Although many of the existing applications are currently less than comprehensive, the large potential revenues from a rapidly growing deployment of distributed Unix client/server environments are tempting these and other vendors to develop and enhance their products and the levels of integration they support.

A number of general-purpose products monitor and manage Unix servers. Practically all of them can be used to manage special-purpose Web servers. Some of the better known products are (Terplan, 1995):

- MaestroVision from Calypso Software
- EcoTools from Compuware
- GlancePlus from Hewlett Packard
- PerfView from Hewlett Packard
- TME 10 for Unix from IBM/Tivoli
- Patrol Knowledge module for Unix from BMC
- TNG Unicenter from Computer Associates

Unix server management requires that multiple information sources be coordinated and correlated with each other. Applications providing information are usually from different vendors. The most common application programs include:

- Hardware monitoring
- Leading management platform
- Log analyzers
- Schedulers
- Print managers
- Back-up managers

7.3.1 PERFVIEW FROM HEWLETT PACKARD

PerfView is a centralized performance measurement tool originally designed to monitor HP operating systems. MeasureWare is a multifaceted product that brings PerfView-collected data onto the same console as data that measures the performance of networks, applications, and databases. MeasureWare includes an end user interface component that can display information collected by selected HP and non-HP agents. One of these agents is the original PerfView agent with some important enhancements.

PerfView consists of two components:

- Motif-based performance analysis software, supporting alarm monitoring, filtering, and analysis capabilities running on HP systems
- Intelligent agents that collect and monitor metrics on HP, Sun, and IBM systems

Agents capture and log statistics, and determine if exception conditions exist. Whereas IT/Operations (IT/O) agents trigger alarms based on specific messages, PerfView agents are more sophisticated and can apply algorithms to compare current service levels, including response times, transaction rates, resource utilizations, and bottleneck indicators against predefined alarm thresholds. PerfView agents are capable of tracking approximately 30 Unix metrics. Metrics include, but are not limited to:

- Total CPU utilization
- Active processes
- Peak disk utilization
- Physical disk I/O per second
- Memory utilization
- Memory management disk I/O per second
- Swap space utilization
- Packet I/O per second
- Cache hit rate
- User CPU utilization
- Active processes

The PerfView management console application includes three components: an analyzer, a monitor, and a planner. The agents send raw performance data via remote procedure calls (RPCs) from the alarm source, which is the managed node, to the analyzer for further analysis. The PerfView console application then filters out irrelevant performance data, time stamping and logging the relevant data in the system's memory. The monitor supports management-by-exception through centralized event monitoring and customized alarms. Alarm conditions can consist of thresholds and time duration elements for single measurements or combinations of multiple measurements. The monitor displays alarm conditions, while the planner provides linear forecasting models based on historical data provided by the agents.

Both MeasureWare and PerfView console applications can manipulate data from non-MeasureWare agents through an extensibility feature called data source integration (DSI). The DSI interface allows operators to monitor any resource, including MVS systems or the output of existing Unix scripts, through a proxy agent. Agents can automatically initiate local actions when thresholds are crossed. These actions include pages, e-mail, scripts, or other programmatic responses.

7.3.2 GlancePlus from Hewlett Packard

This is an online diagnostic software product that supports examination of performance data at the Unix system application and preprocess levels. Problems with system CPU, memory, disk, and network utilization can be detected using the product. Information can be displayed in graphical or tabular form. GlancePlus supports user-specified performance rules for diagnosing problems; alarms can be triggered if problems are detected. In addition to supporting metrics on individual processes, GlancePlus allows several measurements to be combined and evaluated as one metric. GlancePlus graphs typically are used to show the short-term performance history of cumulative and running average totals over the course of between 10 minutes and several hours. GlancePlus includes a rules-based advisory facility. The product is capable of reporting on more than 600 system metrics.

7.3.3 PERFVIEW RX FROM HEWLETT PACKARD

This product is an add-on application that can be used for long-term trending and capacity planning. PerfView RX can track more than 200 system statistics. It identifies applications and processes using Unix system resources. A number of metrics can be displayed, including the login ID of each application or process. It allows concurrent graphing of multiple metrics such as total CPU utilization, peak disk utilization, disk I/O, etc. The display helps users spot the cause of system bottlenecks via flexible graphing options. Users can turn on or off the lines representing various metrics. The granularity of the graph can be adjusted for viewing of any trends or anomalies in system activity. Users can adjust the axis of the graph to more closely view lower level activities. Metrics displayed by PerfView RX include, but are not limited to the following:

- CPU use during interval
- Number and rate of physical disk I/Os
- Maximum percentage of all disk file sets
- User CPU use during interval
- CPU use at real-time priorities
- CPU use for interrupt handling
- CPU idle time during interval
- CPU use for managing main memory contents
- Rate of system procedure calls during interval
- Number of disk drives configured on the system
- Average utilization of busiest disk during interval
- Number and rate of logical/physical disk reads, I/Os, writes, etc., during interval
- Percent of logical reads satisfied by memory cache
- Number of configured LAN interfaces
- Number and rate of network file system requests during interval
- Main memory use as percentage of total
- Number and rate of memory page faults during interval
- Number of process swaps during interval
- Percent of virtual memory currently in active use
- Number of processing in run queue during interval
- Number of user sessions during interval

The reporting facility is very useful for consolidated reporting for multiple information sources that are not limited to Unix-related metrics.

7.3.4 PATROL KNOWLEDGE MODULE FOR UNIX

Most applications, databases, Internet technologies, and middleware systems running in enterprises are mission-critical. For each of these technology components to be available and running at peak performance, the Unix system they run on must also be available. In today's rapidly changing and fast growing technology environments, it can be challenging to ensure all components, including the important Unix Web servers, are always available for users. In addition, with changing Unix technology and subtle differences in Unix from other vendors, it is difficult to be familiar with the nuances of every system. A tool to help administrators automatically monitor and manage their diverse Unix systems is required.

The Patrol Knowledge Module for Unix automatically monitors and manages Unix systems and related resources. Patrol provides current and historical information through a centralized console. Users can easily see the status of the Unix environment. From high-level areas such as CPU, file systems, and printers, users can drill down to view detailed parameter data. For example, in the file system area, Patrol tracks:

- Storage capacity
- Available space
- Number of unused i-nodes
- Percentage of i-nodes in use

In addition to monitoring parameters, Patrol can proactively take corrective actions when things go wrong or are about to go wrong. Through this functionality and more, Patrol helps users quickly determine availability and ensure performance for Unix servers. Monitored parameters include:

- Active processes (e.g., ranks active processes consuming the most CPU time or using the most memory)
- CPU usage (e.g., identifies activities that may degrade system performance and impact user productivity)
- Disks and disk usage (e.g., tracks disk I/Os and usage to locate hot [overused]) disks)
- File system resources (e.g., determines disks that are heavily used and running out of space)
- Kernel processes and kernel resource usage (e.g., monitors i-nodes, the file table, and the process table)
- Log files (e.g., specify the log files to monitor and set alarms based on the size or growth rate of the log file or the occurrence of user-specified text strings in the file)
- Memory activity (e.g., monitors paging, I/O caching, and swapping)
- Network activity (e.g., monitors traffic levels on TCP/IP, measuring the ongoing health of the network by developing benchmark values useful in analyzing overall trends in usage)
- Remote procedure call and network file service activities
- Print queue and printer activity (e.g., monitors the size of various print spools and provides alerts if any become full)
- Active and zombie processes (e.g., run process-oriented commands such as listing zombie processes)
- System swap files (e.g., monitors the percentage of swap space used and free)
- User sessions and processes (e.g., tracks users on the system and identifies which users and processes are placing the greatest demands on system resources)

Key features of the Patrol Knowledge Module for Unix are:

- Obtains monitoring information from both OS and kernel levels
- Allows users to execute Unix commands through menu items
- Provides information through console views, InfoBoxes, reports, graphs, charts, and gauges

Key benefits of the Patrol Knowledge Module for Unix are:

- Delivers automated resource monitoring and management functions
- Executes proactive and automated corrective actions to solve many problems and potential problems
- Supports optimum performance, reliability, and availability for related system resources

7.4   MONITORING NT WEB SERVERS

NT server management includes in-depth monitoring, analysis, and control of the performance, fault, and configuration aspects of the server operation. Effective NT server management requires:

- Built-in capabilities to the server and options
- Strict adherence to popular network management standards for communication protocols and data interchange
- Tools that operate within the customer's preferred environment.

The combination of these capabilities allows the customer to manage the NT server to ensure that it is one of the most reliable resources in the network.

NT server management requires that multiple information sources be coordinated and correlated with each other. Applications providing information are usually from different vendors. The most common application programs include (Figure 7.3):

- Insight Manager from Compaq
- Backup Exec from Seagate
- SMS from Microsoft
- MMC (Microsoft Management Console) from Microsoft
- NT event logs for systems, applications, and security from Microsoft
- Performance monitors from Microsoft
- Schedulers
- Print managers

These information sources represent different formats and viewing capabilities. Most of them support de-facto management standards, such as SNMP, MIB, DMI, and MIF.

A number of general purpose products monitor and manage NT servers. Practically all of them can be used to manage special purpose Web servers. The better known products are:

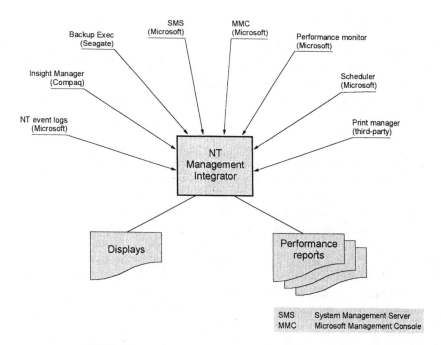

FIGURE 7.3   Information sources for NT management.

- RoboMon from Heroix
- ManageX from Hewlett Packard
- TNG Unicenter from Computer Associates
- AppManager from NetIQ
- Netune Pro from BMC
- Manage Exec from Seagate
- TME for NT from Tivoli

A few typical examples from this product list are provided here.

7.4.1   RoboMon from Heroix

RoboMon provides centralized management of distributed NT servers. However, it must be installed at each managed server by manually logging onto the server executing its installation program. As an alternative, the installation program can be deployed as an SMS package. To perform the installation, the user must have administrative rights on the server. RoboMon's installation program has three options:

- Full — supports management and monitoring of servers
- Agent — installs the agent to allow a server to be managed
- Console — installs the management console to monitor servers

Regarding security within the application, Heroix states that permissions can be set on a specific file to limit who has access to apply rules on a server. However, users with administrative rights can also "take ownership" and assign themselves access to the appropriate files.

Managing servers — Once RoboMon is installed on a server, the RoboMon enterprise manager can be brought up. This is RoboMon's management console and will allow the user to invoke RoboMon "rules," monitor alerts, and perform reporting functions. RoboMon includes a large number of rules "out of the box." However, if a process is identified that is not supplied with RoboMon, a rule may be created. The programming language within RoboMon is proprietary but fairly simple.

Alerts — When monitoring an NT server, one must configure the destination of alert messages. Servers can be configured to send messages to multiple consoles simultaneously. RoboMon can also be configured such that a management console can send an alert to another RoboMon management console. To receive these messages, the management console must be powered on. However, the enterprise management application is not required to be running. In the event the management console is not powered on, RoboMon provides guaranteed message delivery by attempting to resend the message for a user-definable duration.

Historical reporting and trending analysis — By default, 32 days of historical data are stored on the managed server's local hard drives in a Microsoft Jet database. If a centralized approach to maintaining historical information is preferred, this can be configured as well. However, this is a manual process that must be configured by the RoboMon administrator.

Miscellaneous — RoboMon has been on the market the longest of the NT management applications. It is also capable of monitoring various host-based systems.

Summary — RoboMon is a flexible tool to manage NT servers in a decentralized fashion. Because of its flexibility, it is feasible that management of the product could become an operational burden. External to this issue, RoboMon has three major architectural weaknesses, which are:

- RoboMon must be deployed either by physically logging on to the server and running the setup program or by deploying an SMS package.
- After the product is installed, the server will most likely need to be rebooted before it can be managed.
- The programming language used for creating custom scripts is proprietary.

Due to this architecture, implementation is likely to be labor intensive, resulting in a higher deployment cost.

7.4.2 ManageX from Hewlett Packard

ManageX is a DCOM-based application and is based on the initial release of Microsoft's Management Console (MMC). Because it is based on DCOM, its minimum operating system requirement is NT 4.0. Deployment begins with the installation of the application on a system. From within the MMC, a ManageX "policy" is pushed out to a selected NT server which will enable the server to be managed.

By default, ManageX policies are maintained on the local system. This function can be manually configured to share policies across the network. However, security within the application is based on NT file-level security. As a result, once the software is installed, there is no way to prevent deployment of policies.

Managing servers — Once the agents are installed on a server, ManageX policies may be centrally deployed to monitor objects within the server. Upon invoking the ManageX console, two additional windows — device manager and command queue — are opened. Now, the administrator will add the targeted servers to the device manager window, select the servers, and deploy the ManageX policies to the servers accordingly. ManageX's supply of policies "out of the box" is very strong. However, in the event an object or process is not included in its standard supply of policies, the programming language is ActiveX based.

Alerts — When configuring the ManageX agents on a managed server, the administrator must also determine where alerts should be sent. By default, all alerts are broadcast on the local NT domain as well as the local subnet. All systems running the ManageX agent will receive the alert. To modify this process, the administrator must create and deploy a system route table (SRT) policy. The SRT specifies the destination for all alerts generated on the managed server(s). These alerts may optionally be written into a Jet database. Although simple to configure, this is not the default and therefore it must be manually set.

A second alerting function of ManageX is the message routing policies (MRPs). MRPs provide a forwarding function for alerts received by a system as specified in deployed SRTs. Together, the functions of the SRT and MRP provide the ability to create a cascading alerting mechanism.

Integration between OpenView's IT/O and ManageX is very important from an operational viewpoint. It provides the functionality referred to by HP as "lights out console." This functionality is accomplished by installing the IT/O agents for NT and configuring the agent to forward all messages — those received by both SRTs and MRPs — to the IT/O console. However, the communication between the lights out console and the IT/O console is unidirectional, because the IT/O console can neither acknowledge nor update events generated by ManageX.

ManageX currently provides a Web front-end for viewing the current status and alerts of managed devices. This interface is currently read-only with the exception of acknowledging alerts. This limited functionality is common across all products. This Web back-end also provides a method for an IT/O console to acknowledge events.

Historical reporting and trending analysis — A lot of similarities can be observed with Unix performance reporting. One of the major enhancements in ManageX is the addition of recording performance data for the purpose of historical reporting. By default, each managed server will maintain its own information locally in a Microsoft Jet database. ManageX will provide a method of forwarding the recorded performance information into one or more centralized database(s). The centralized database can be Jet, SQL Server, or any ODBC-compliant database. This database can then forward its data into another database, thus creating a cascading effect. As the centralized database grows, HP recommends it be maintained within an SQL Server database. Currently, HP is unable to provide recommendations regarding sizing and when the centralized database should be converted to an SQL Server database. When the centralized database becomes SQL Server based, the total cost for implementing ManageX will increase because both hardware and software costs will increase.

Summary — ManageX is a relatively new product under the control of Hewlett Packard. It needs some time to be fully integrated into management applications, such as IT/Operations and IT/Administration (IT/A). Combined with both IT/O and IT/A, HP could offer a seamless suite of management applications for both Unix and NT environments.

7.4.3 APPMANAGER FROM NETIQ

AppManager is a client/server application based on SQL Server and COM. The two main components of the application are the management server and repository server. These are logical components that may be maintained within one or two physical servers. Installation of the management and repository servers (MS/QDB pair) requires NT 4.0 and SQL Server 6.5 with Service Pack 3 or higher.

During installation of the management and repository server components, the administrative software can be installed on the same system. The same administrative software can be installed on an NT workstation — this is very similar to installing the SQL Server administration tools. By invoking the administrative software, the application requires the user to connect to the SQL Server (repository server). By connecting to an SQL Server application, AppManager is able to provide an additional layer of security external to NT file-level security. Upon establishing the connection, the administrator can add servers to the device list and begin deployment. Monitoring parameters within a server is accomplished by deploying knowledge scripts (KS). When a KS is deployed, it is recorded into the repository server's database and is then automatically pushed out by the management server.

Microsoft provides Systems Management Server (SMS), but it focuses primarily on deployment issues such as software distribution and not on critical operational issues such as event and performance management. Microsoft also provides various stand-alone tools that come with Windows NT and each BackOffice server, but these tools are not built to manage a distributed environment.

The AppManager suite is comprised of integrated AppManager components that are specifically designed to monitor the performance and availability of Windows NT-based systems and Microsoft BackOffice servers such as Exchange Server and SQL Server as well as Lotus Domino/Notes servers. Using the AppManager console, administrators can configure KS monitoring functions to collect performance data and monitor for simple or complex events. Corrective actions — such as sending e-mail, generating an SNMP trap to a network manager, or running a corrective program — can also be easily set up to automatically execute when a specific event occurs. The result is a powerful and automated closed-loop solution for both pro-active problem detection and problem resolution across a customer's highly distributed Windows NT environment.

The AppManager Wbem agent is the architectural component that runs the KSs on Windows NT servers and workstations. It implements the Wbem technology

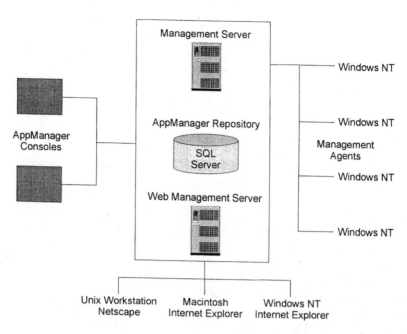

FIGURE 7.4   Architecture of AppManager.

architecture as a primary method of accessing instrumentation data and also makes its data available to other management products that support Wbem. With Windows NT 5.0, Wbem agents are automatically included.

Besides providing much needed operations management capabilities for Windows NT and Microsoft BackOffice environments, AppManager also lowers the total cost of ownership of these systems and applications by letting customers leverage scarce and valuable personnel resources. AppManager accomplishes this by automating repetitive and time-consuming tasks, providing prepackaged knowledge and business rules for managing and monitoring distributed Microsoft BackOffice environments, centralizing management of distributed and remote systems and applications, and enabling proactive notification and correction of problems before they impact the customer's business.

To support scalability, NetIQ provides a very flexible architecture comprised of the following components (Figure 7.4):

- Console — A user-friendly program that represents systems and applications as resource icons; the console is used to centrally define and control the execution of KSs on the managed NT systems and applications via simple drag-and-drop actions
- Repository server — A central repository based on Microsoft SQL Server that stores management data
- Management server — A service that runs on a Windows NT server that manages the event-driven communication between the repository and the NetIQ agents

- Agent — A highly intelligent program that runs on any Windows NT system. The agent receives requests from the management server to either run or stop a KS, and in turn communicates back on an exception basis any relevant data or events collected by the running KSs
- Web management server — A set of active server pages that lets the user monitor the entire Windows NT environment from a Web browser

This robust, multi-tier architecture makes AppManager a scalable foundation for managing small- to large-scale deployment of Windows NT. AppManager is scalable because processing is spread across multiple architectural components, enabling the management server to control large numbers of agents. Push technology may also be used to distribute AppManager agents to remote systems for easy deployment. This architecture uses an exception-based mechanism to communicate between its components, so no polling occurs to unnecessarily consume precious network bandwidth. This architecture is also reliable because its components have been built for continuous and autonomous operations.

The NetIQ AppManager Wbem agent consumes and provides Wbem-based management instrumentation data via the CIM, a key part of the Wbem technologies. The CIM is a method for storing management and instrumentation data in a common format. Because AppManager Wbem agent's data is available in the CIM format, the system is compatible with other Wbem-enabled management tools.

Managing servers — To manage an NT server, the user must have administrative rights on the managed server plus proper access within the management server. Deployment of AppManager begins by pushing out the "general agent install" KS. After this is completed, additional KSs may be deployed to monitor specific objects and processes within the server. In general, installation of the agent on a server does not require a reboot. However, there are two instances where a reboot is required. Initial installation of the SMS management agent requires a reboot because the location of the SMS application must be appended to the system's path and NT cannot currently do this dynamically.

NetIQ includes a solid number of KSs with their product. However, if an item needs to be monitored that is not included with the product, additional scripts may be created. The programming language used by AppManager is VBA (visual basic for applications).

When a KS is deployed to a server, the user can define the interval as well as the desired schedule. Another option is the ability to specify whether the object's status will be logged back into the repository server's SQL Server database. NetIQ supplies a strong selection of KSs for all of the Microsoft BackOffice products.

Alerts — All alerts are sent to the repository server's database. If the database is unavailable, the information is stored locally in a Jet database. When the managed server is able to reestablish a connection, the information is uploaded to the repository server. These messages are viewed and managed via the repository and management server. As a result, there is no management of alert message routing. However, the application is dependent on the MS/QDB pair being online.

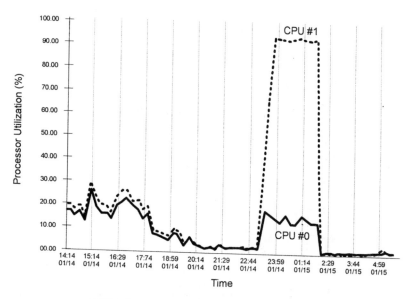

FIGURE 7.5 CPU utilization by AppManager.

Due to the architecture of the product, AppManager supports the ability to assign permissions to users such that they have the ability to view, acknowledge, and close alerts. The same users can also enter comments into the database, detailing the action(s) taken as a result of the alert. However, users would not be able to deploy new KSs.

Integration between IT/O and AppManager is dependent on the IT/O agent for NT. This agent provides the ability to send a message to the IT/O console. However, this communication is unidirectional because the IT/O console is unable to send a message back to AppManager. Optionally, the IT/O console can acknowledge alerts via a Web browser. This is common for all applications tested.

Historical reporting and trending analysis — When a KS is deployed to a server, one of the options is to log the resulting value(s) of the KS into the repository server. If enabled, at the interval defined by the KS, the information will be logged. The amount of data that is saved is limited to the amount of storage space on the repository (SQL) server. When historical data is needed, the user must connect to the repository server and generate the selected reports. NetIQ provides an ample supply of standard reports within the application. AppManager also supports the creation of custom reports using Crystal Reports. These reports can also be published as a Web page for quick access. Figure 7.5 shows the utilization of the CPUs; Figure 7.6 displays I/O activities over the same time frame.

NetIQ does not provide paging functionality within the application. However, AppManager does provide integration with a solid number of paging vendors.

Summary — AppManager is a new product with a lot of potential. The tight collaboration with Microsoft helps to integrate the product with other MS management

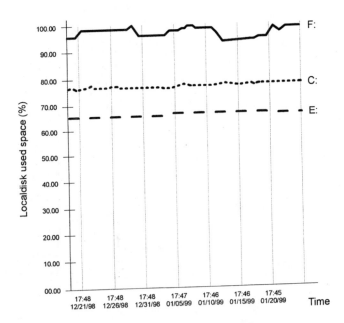

FIGURE 7.6   I/O utilization by AppManager.

tools. It supports the new Wbem standard, resulting in realistic opportunities for unified enterprise management.

7.4.4 MANAGE EXEC FROM SEAGATE

This product tracks more than 1000 attributes of NT systems. It provides an instantaneous picture of a server's health, as well as historical and trend data. Manage Exec supports forwarding of SNMP traps to OpenView, TME 10, Unicenter TNG, and NerveCenter from Seagate. It also plugs into MMC. Following the industry trends, it supports a Web console and a Windows-based console. In addition to the consoles, which are user interfaces for the system, Manage Exec consists of two other component groups (Gibbs, 1998):

- The executives, which receive alerts and decide how they should be handled
- The agents, which monitor server information and send out the alerts

Manage Exec deals with events, which simply are changes in system performance variables. Depending on how event handling is configured, events may be assigned to one of five levels of severity. Events can be defined for when counters go above or below certain values and when they change by a set amount.

Seagate uses two kinds of executives: a Web executive, which provides the Web interface for browsers (essentially a customized Web server front-ending the system), and an alert executive, which offers up views of alert data from the agents. The alert executive and alert server run as system services under Windows NT.

There also are two types of agents: monitoring agents which collect real-time server data, and alerting agents which periodically scan for server conditions that exceed preset thresholds. Agents are implemented as services under Windows NT. Agents have little impact on the servers on which they run. They are likewise designed to minimize network traffic by exchanging statistical information during scheduled bulk transfers, rather then being constantly polled by the executive. The agents track counters within the monitored operating system. These counters include network and disk I/O and processor utilization. Events are sent when counter values change or reach predefined minimum or maximum values. Events can be created by counter changes or by threshold violations. In response to events, the Windows console can be configured to make sounds, flash its icon in the task bar, and activate pagers.

Through the Windows console, the user can examine current alerts and clear outstanding ones, as well as browse historical events. Obviously, at large numbers of monitored NT servers, a considerable number of events can be generated and transmitted that may overload facilities. For such cases, Manage Exec provides filtering capabilities. Many of the reports can be accessed by Web browsers.

There are many configuration alternatives in use: management and monitoring single NT servers by one or more consoles, or monitoring multiple NT servers by one console or by multiple consoles. Overhead and internal performance metrics may be controlled by the number of monitored counters, how often counters are evaluated, and how long statistics are retained.

7.4.5 PATROL KNOWLEDGE MODULE FOR NT

To effectively manage the applications that drive businesses, users need to manage all the components on which the application relies — the database, Internet technology, middleware, and the operating system. Managing these heterogeneous components can be challenging, especially as the number of each component grows.

Patrol uses distributed intelligent autonomous agents and loadable libraries of expertise called knowledge modules (KMs) to automatically discover the environment, continuously survey related systems, implement recovery actions, and initiate alarms based on preset parameters. By proactively monitoring and managing complex environments, Patrol helps ensure performance and availability.

Patrol KM for Windows NT contains the knowledge used by the Patrol agent for NT monitoring, analysis, and management activities. Patrol statistics are collected at the same point in time and stored, which allows for comparative analysis of performance trends over time. Patrol continuously monitors statistics, or NT parameters, to ensure that systems are running at peak performance and can obtain the same data as the NT performance monitor. This product also provides a comprehensive list of menus as commands to make the administration of NT easier.

Monitored NT parameters include:

- Cache — monitors memory file pages and activity such as the frequency of cache faults per second
- CPU — monitors CPU usage and activity such as number of device interrupts
- Event logs — monitors system, security, and application event logs and lets users view, back-up, or clear
- Logical disks — monitors and provides file system resource data such as the ratio of the free space available on the logical disk unit to the total usable space
- Memory — displays memory statistics such as number of available bytes and size of paged and nonpaged memory pools
- Network — monitors network activity and provides information about the rates at which bytes and packets are received and sent over a TCP/IP connection
- Pagefile — monitors memory swapping and paging activities such as peak and pagefile use
- Physical disks — monitors disk resources
- Printers — monitors local printers or printers associated with the system
- Process — monitors process threads and virtual memory usage
- Security — monitors network and server security such as checking the number of times an attempt to open a file has failed due to no authorization
- Server — monitors server processes
- Services — monitors for the presence of a service and optionally restarts if the service has stopped
- System — monitors the state of the NT objects and system to make sure capacity is sufficient for current demand

Key features of the product are:

- Proactively manages all critical aspects of Windows NT
- Consistent and continuous monitoring
- Historical record of events and statistics
- Repository of expertise
- Extensible to provide application monitoring

Implementing and using this product offers the following benefits:

- It delivers automated resource monitoring and management functions.
- It allows system administrators to cross-leverage their expertise into their Windows NT environment.
- It supports optimum performance, reliability, and availability for related system resources.

7.5 UNICENTER TNG WEB MANAGEMENT OPTIONS (WMO)

The Unicenter Web Management Options (WMO) enable comprehensive management of intranet, extranet, and Internet Web sites to make business processing reliable, efficient, and secure. Agents provide end-to-end management of the Web environment, timely notification of error conditions and performance degradation, and automated monitoring of all Web elements. WMO makes it possible to manage Web site health and performance on an enterprise scale. It helps to control the total cost of ownership and improve the level of service by providing the following:

- A comprehensive model for integrated Web site management
- Site monitoring with optional automatic corrective actions
- Centralized event correlation and policy-based management
- Correlation of site health and performance from both end user and server perspectives
- Correlation of Web site performance with Web server OS and network response
- Real-time and historical reporting across multiple sites

There are two key components of WMO: the Web response agent and the Web server agent. Both can be deployed in both Unix and NT servers. The response agent provides automated, enterprise-wide health and performance monitoring from the user or universal Web browser perspective. The Web server agents are intelligent, programmable agents that monitor the health and performance of a Web server site. Events from the Web response agent can be correlated with information reported by the Web server agent. In addition, other Unicenter TNG agents, such as the system agent (on which the Web server is running), network response option agent, database agents, and Microsoft's transaction server agent, can be correlated to provide truly comprehensive end-to-end management (Sturm, 1998).

7.5.1 WEB RESPONSE AGENTS

Web response agents remotely monitor the accessibility of Web servers. This allows the site administrator to immediately detect and respond to server and network performance problems. The agent impersonates a Web browser by accessing the Web server from a remote Unicenter TNG server and periodically reports availability based on the response time of the server.

Web response agents can reside on one or more geographically distributed systems and can use standard Internet protocols to monitor the health and performance from the user perspective. The agent uses a combination of Ping, DNS protocol, FTP, HTTP, and HTTPS to determine the availability, roundtrip response, and content of select URLs. The agent saves the route used to access the service as a string. The agent also automatically correlates the status of Ping, DNS, HTTP, and HTTPS for better problem resolution.

The features that apply to the overall operation of the Web response agent follow:

- FTP: The administrator can monitor URLs with an FTP scheme, such as ftp://ftp.cai.com/readme.txt. The agent will attempt to use FTP to copy the file and compute the delta time from when the request was made, when the first byte was received, and when the last byte was received. The administrator can optionally specify a CRC to verify the content of the file.
- HTTP: The administrator can monitor URLs with an HTTP scheme, such as http://www.cai.com/. The agent will attempt to use HTTP to retrieve the URL and compute the delta time from when the request was made, when the first byte was received, and when the last byte was received. The administrator can specify a string or CRC to verify the contents of the URL. The agent can retrieve all the components of the URL (page) or just the main body. Any components of the page that cannot be retrieved will result in an event report and content error status. The administrator can configure the number of concurrent threads to create when retrieving a single URL and the maximum number of URLs the agent will attempt to retrieve simultaneously. All chaining is disabled, and the agent computes the total number of bytes received and the average throughput. CGI and ASP (Active Server Pages) are handled just like HTML.
- Proxy: The agent supports proxy configurations on a per-URL basis.
- Diagnosis: The agent issues a Ping and/or DNS test to diagnose service errors automatically. The administrator can control the operation for each status: normal, warning, and critical. For example:
  Normal — URL
  Warning — Ping, URL
  Critical — Ping, URL, DNS
  In this example, the agent will normally access the URL. In critical status, the agent will issue a ping test and a DNS test and will attempt to access the URL.
- Correlation: The agent automatically correlates Ping, DNS, FTP, and HTTP status. For each URL monitored, the Ping response, DNS response and accuracy, and FTP and HTTP response and accuracy are displayed.
- Historical reporting: The default configuration for historical data collection is the URL name, the current and average time it takes to receive the first byte and last byte, and the total number of bytes transferred.

7.5.2 WEB SERVER AGENTS

Web server agents monitor the Web server for the Webmaster. The Webmaster can use the agent to monitor the same set of URLs as the Web response agent. Events from the Web response agent can be automatically correlated with information reported by the Web server agent. This also allows the Webmaster to engineer performance by correlating roundtrip response time as perceived by the end user with performance metrics generated by the Web servers. In addition, information from other Unicenter TNG agents, such as the system agent (on which the Web server is running), network response option agent, database agents, and Microsoft's transaction server agent can be correlated to provide truly comprehensive end-to-end management (Sturm, 1998).

The Web server agents are intelligent, programmable agents that can monitor the health and performance of a Web server site. Web server agent supports various platforms:

- Microsoft IIS
- Netscape Enterprise Server
- Netscape FastTrack Server
- Apache Web Server
- Sun Java Server
- Lotus Domino Web

The agents implement a consistent model for the monitoring and management of Web servers. The features that generally apply to the overall operation of the Web server agents are:

- Status: The agent maintains and reports three levels of status: normal, warning, and critical. Status is summarized at the component instance level and is propagated up to the component container and to the agent (top level). The administrator can easily determine the overall state of the agent by examining a single attribute. This propagation provides for progressive disclosure when drilling down to isolate the source of a problem. Summary status counts are maintained on a component basis, including the number of instances being monitored and the number that have warning and critical status.
- Reset: The administrator can reset the alert counts and statistics on demand. In addition, an automatic reset interval can be specified. The agent will reset the values to zero automatically at the specified interval. An interval of zero disables the automated reset.
- Statistics and availability: The agent computes the current response time, high, low, rolling average, and standard deviation. In addition, the agent computes the availability of a service.
- Thresholds/actions: The administrator can define warning and critical thresholds. A warning threshold must have a value that is less than or equal to the critical threshold. This is enforced by the agent. A warning or critical threshold can be enabled or disabled. Corresponding to each threshold, a local action can be specified. A local action is simply a command line that is executed within the context of the system on which the agent is running. An audit log maintained by the agent includes the command line, a time stamp, and status. If the command line fails, an event is reported. Regardless of the status, an event is reported to indicate that the local action was taken, which provides a convenient audit on the manager.

- Services: The agent can automatically monitor and restart stopped services/deamons, such as HTTP and FTP, and can optionally reboot the Web server after "n" failed restart attempts.
- Counters: The agent can be customized to monitor a select set of counters, which indicate the performance of the Web server.
- Disk space: The agent monitors work space, log files, and Web content disks for available free disk space.
- FTP: The Webmaster can monitor select URLs with an FTP scheme, such as ftp://ftp.cai.com/readme.txt. The agent keeps real-time statistics and records the number of hits, invalid logins, IP disconnects, and response time, including the preceding statistics.
- HTTP: The Webmaster can monitor select URLs with an HTTP scheme, such as HTTP://www.cai.com/. The agent keeps real-time statistics and records the number of hits, invalid logins, HTTP errors, IP disconnects, and response time, including the preceding statistics. The agent records the peak number of simultaneous hits within a specified interval, such as 100 hits within 15 seconds. The agent can also be configured to reject a request if the number of simultaneous hits exceeds a threshold.
- Web crawler: The agent uses a low-priority thread to crawl all of the links on the Web site, and reports any errors (e.g., bad links) into a table. Real-time HTTP errors are reported in the table as well.
- Event notification: All errors and status changes result in an event report (SNMP trap) to the manager, which in turn reports messages to the event manager component of TNG. This combination permits event correlation across objects and systems. Event manager includes a number of methods for programmed event action.
- Aggressive retry: When warning or critical status is detected, the agent will automatically shorten the monitor interval, in effect performing an aggressive retry. The agent automatically returns to the regular schedule when status is normal. Three monitor intervals can be specified for each row in the table: normal, warning, and critical.
- Correlation: The Webmaster can define a cluster of Web servers, which makes up a Web site. A Web site container can be defined as well. Each Web server is managed separately, but the status is propagated to include the whole Web site.

Unicenter TNG is altogether a powerful choice. In particular, enterprises where both Unix and NT Web servers must be managed together can benefit from this product. Its use enables businesses to confidently exploit the Internet and intranets to their full extent. Through comprehensive, integrated Web server management, this option

- Secures business information through policy-based user authentication and authorization — The Internet and intranets expose businesses to new security risks. TNG Unicenter supports existing secured communication capabilities and provides the authorization and access control that existing

Web server software lacks. It also protects HTML pages based on user authentication and authorization, secures HTTP proxy servers, and provides full audit trails of Web page accesses and attempted security breaches. This allows customers to conduct business through Web sites, while protecting the contents of HTML pages.

- Conserves disk space by automatically archiving HTML pages and other Web site objects — This Web management option monitors how frequently Web pages are viewed and automatically offloads infrequently used HTML pages. An intelligent restore capability transparently brings them back when accessed. This hierarchical storage management conserves disk space while preserving links to HTML pages.

- Improves Web server availability by automatically detecting and correcting errors — Status of Web servers is continuously supervised. There are built-in policies for how to react automatically to events and logs. The product correlates system, network, database, application, and Web server events, providing a complete picture of the state of the Web site. In particular, memory, CPU, and file usage are monitored. Furthermore, the product can detect abnormal file size growth, disk space utilization, and file accesses. Administrators can analyze patterns of usage and peak consumption to help improve effectivity.

- Improves service levels by providing end-to-end performance management — Customers accessing Web sites demand immediate response times. If they are unable to access a Web site, they go elsewhere. The TNG Internet option continually monitors Web server performance — such as response times and queue lengths — and displays them in the TNG console. Using filtering and correlation policies, agents trigger events to notify administrators of potential problems, open trouble tickets, or automate actions.

TNG is one product that is able to integrate both Unix and NT server management.

7.6 OPENVIEW FROM HEWLETT PACKARD

The new HP OpenView Internet service manager software enables HP OpenView to collect performance metrics for the Netscape Web server and Microsoft's Internet information servers — data that is useful for proactive intranet planning and investment decisions. It enables OpenView to manage all standard services for Windows NT functionality. OpenView's Internet service manager also has a link-auditing subsystem for content management, which ensures uninterrupted access to a company's online content. The present offer includes management and security integration, intuitive browser GUIs, and management and planning tools.

To address the issue of Internet security, HP has partnered with Check Point, Raptor, and TIS to allow mapping, polling, and management of network devices through the companies' respective firewalls without compromising corporate security. In addition, OpenView network traffic now can be encrypted, preventing unauthorized access to potentially mission-critical data.

All Internet and intranet elements can be managed by a single view, significantly improving problem isolation, diagnosis, and resolution. HP's resource and performance management solutions build on its consolidated view to help IP managers measure, monitor, and manage Web server resource utilization and performance. This functionality allows them to proactively maintain and improve service levels on an intranet before end users are impacted. As stated previously, the majority of existing tools can be combined to support intranet monitoring and management.

7.7   PERFORMANCE OPTIMIZATION

Usually, users find high-traffic Web sites with slow access, unpredictable delays and low-quality page content. There are a few new tools that directly address Web server performance and load balancing issues. They reside in the same server or in a separate server and supervise key performance metrics in real time.

When a Web server load balancer is in place, Web clients connect to a virtual Web site. The load balancer distributes client requests to multiple servers based on traffic volumes, availability, and other criteria. Many of the products also provide content-based balancing, which allows Webmasters and site administrators to match traffic to server power by designating a large, fast disk for image content and a fast and secure server for SSL and electronic commerce transactions (Hoover, 1999). Some load balancers can even isolate Common Gateway Interface requests and send them to a separate server or can prioritize SSL traffic over HTML traffic. More advanced Web server load balancers can operate globally, managing traffic among multiple, geographically dispersed sites. Global load balancing ensures that content and applications are quickly and cost-effectively made available to remote users.

Before evaluating load balancing products, Webmasters and site administrators should characterize their sites (Hoover, 1999):

- They should determine their application requirements, which will determine the type of persistency policies they need. In the Web server world, persistence means a repeat user is connected to a Web site that is aware of that user's previous requests and account status.
- They should determine the scale they need now and estimate the scale they expect to need in the future based on hits per second, number of simultaneous users, and bandwidth requirements. They should make sure that they take into account the maximum capacity of other key elements of their systems, such as WAN bandwidth capacity, database server capacity, and application server capacity.
- They are expected to consider their application architecture. Is it two-tier or N-tier, with different levels of transaction processing at each tier? These answers will affect their monitoring requirements. The more complex their transactions are, the more refined monitoring features are needed to highlight performance bottlenecks.

Web server load balancers use the following criteria to decide which Web server or Web server farm should get the client's request (Reardon, 1999):

- Proximity of the server
- Response time and latency
- Packet loss
- Server availability
- Server load

The most common way to gauge proximity is by counting the number of router hops between the client's local DNS server and the content server. Several vendors allow proximity load balancing to be configured manually, enabling intranet architects to set policies that restrict traffic to particular geographic areas.

Response time and latency may be quantified by issuing pings to measure the delay between a client's local DNS server and Web server sites. Since most products calculate delay in real time, they may have to wait for pings to be answered, which could slow response time in some cases.

The estimation of packet loss also helps to determine the best path. This is an important metric for sites that host streaming video and voice or other multimedia.

Server availability is very important and must be checked in real time. DNS doesn't help with this metric. Issuing pings or connecting to an SNMP manager could help to get the necessary information.

Besides server availability, other statistics on the actual site load should also be evaluated in real time. Servers with the lightest load may be selected for the client's request.

7.7.1 SPEEDSERVER FROM SITARA

SpeedServer sits in front of a company's Web commerce and content servers, providing multiple and simultaneous TCP/IP sessions to download needed content. Typically, browsers today have to establish multiple TCP/IP sessions to access a single home page and catalog item because each page could contain many objects, and each TCP/IP session can access a single object at a time. Although improvements in HTTP will let browsers establish multiple simultaneous connections to access Web information, delays still occur in instances where information is located in different Web servers, requiring separate connections. The SpeedServer collects information from multiple Web servers in a single TCP/IP session. The improved throughput between browsers and Web servers could help businesses upgrade their pages with advanced graphics, video, and image content.

SpeedServer reduces the latency over the Internet by eliminating the traditional TCP/IP method of waiting for acknowledgment of connections before sending actual data. Data is sent even as the TCP/IP session is being set up and if any data is lost, only that particular data is resent, unlike the standard method that requires entire segments to be resent. SpeedServer matches the throughput of the modem on the desktop, preventing overload situations. The solution is shown in Figure 7.7. Typical connections share the following problems:

- Unnecessary handshaking
- Good data are retransmitted with lost data
- Separately retrieved page elements

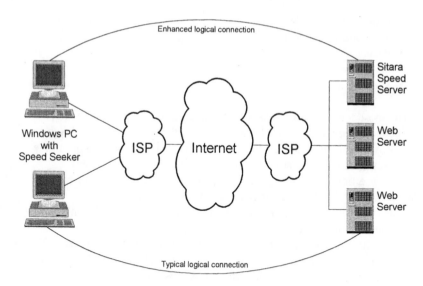

FIGURE 7.7   Enhancing the performance of Web connections by SpeedServer from Sitara.

With SpeedServer enhancements, the attributes are:

- Streamlined handshaking
- Retransmission of lost packets only
- Use of multiplexed data streams

Sitara is introducing a unique client/server application that helps users deliver a significantly improved experience to Web site visitors — which translates into increased revenue, more qualified leads captured, and a deeper relationship with customers.

The Sitara solution is a standards-based application that is easy to install and won't interfere with other network applications. It tackles the root causes of network delay, congestion, and packet loss that make the Web so slow by streamlining the links between Web site servers and a visitor's PC making a page request.

Most existing point solutions (e.g., caching, faster modems, bigger routers and servers, compression, etc.) give users only a limited ability to impact the experience of visitors to individual Web sites. For instance, most servers won't help any sites when the bottleneck is an overloaded router that not the user, but the ISP, controls. All components of an end-to-end link are shown in Figure 7.8. This solution is unique because it is end-to-end, with software adding intelligence on both the client PC and the Web server. The result is that visitors may experience acceptable performance regardless of their modem speed, hardware configuration, or type of connections.

SpeedServer is installed on the same LAN segment of Web servers and can accelerate multiple servers on particular sites. Once SpeedServer is in use, Speed-Seeker can be used as a free download. Web pages are then delivered an average of

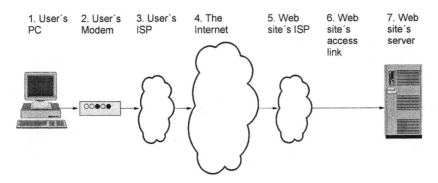

FIGURE 7.8   End-to-end solution with SpeedServer from Sitara.

three times faster, more consistently, and the reliability of the network greatly increases with fewer dropped connections.

SpeedServer is installed on a dedicated server running Sitara's software. It is located on the same LAN segment as the company's Web server site.

The features of the product are:

- Compatible with Web servers supporting HTTP 1.0 and 1.1, including the Netscape Enterprise Server, Microsoft Information Server 3.0, and Apache Server 1.2
- Accelerates all content displayable by Netscape or Microsoft browsers (text, graphic, JavaScript, etc.)
- NT 4.0-based or Unix-based
- No change required to Web servers, except for the addition of one small data file
- Management information includes number of simultaneous Sitara accelerated users, their individual data rate, and the total bandwidth used
- Activity logging in NCSA format

The capacity of a single SpeedServer depends on the contents of the site being served and the connection speed of the visitors being served. For an average Web site, a 300-MHz Pentium-powered server can handle well in excess of 20 Mbps of Sitara-enhanced traffic. Traffic from visitors that are not Sitara-enhanced does not pass through the SpeedServer and thus has no effect on throughput.

7.7.2   WEBSPECTIVE FROM WEBSPECTIVE SOFTWARE, INC.

WebSpective is designed to provide application and traffic management in one solution. It offers high availability and automatic fallover in order to support business-critical Web applications. It also offers content management and centralized monitoring and administration.

The product targets practically all aspects of intranet performance management by watching Web servers, host machines, network interfaces, and Web applications, monitoring their load and performance metrics in real time. The monitoring is performed from the perspective of customers and thus can detect all impacts on business applications due to Web server delays and other delays in the back-end processing. Real-time information may be displayed in configurable graphs, and the system can be set up to immediately notify an operator when certain alarming events are generated.

WebSpective distributes traffic based on the applications that are being accessed, allowing administrators to differentiate the quality of service based on the applications's importance. Furthermore, load balancing among multiple servers is based on real-time application performance metrics, so customers are always directed to the least-loaded and most available server in the farm for their specific application. Failed Web servers can be restarted, and traffic can be rerouted, if necessary.

WebSpective can also prevent problems from occurring by predicting failures and taking preventative measures on servers that are exhibiting certain symptoms. It alerts Web site administrators to high-risk or failure conditions and can automatically off-load traffic to prevent failures from affecting services.

Using its SiteStats Manager, it centralizes the collection and logging of traffic and performance data for an enterprise's Web applications, Web servers, and host machines. The data from all Web servers over any geographical area may be centralized in a single relational database for flexible reporting and analysis. This enables administrators to see historical trends and to perform capacity planning. It also enables them to generate reports using third-party reporting tools such as Crystal Reports or SAS Analytical System Reports.

This product provides a solution to the problem of content management when multiple applications run across mirrored servers throughout a geographically dispersed, perhaps worldwide, Web site. It can do this without causing service interruptions or customer service delays during the update process. Administrators can maintain control over how, when, and where content is distributed, while WebSpective automatically and synchronously posts secure updates. It does this while it maintains state for all connected users, so that end users see no disruption. End users are always served the correct content, even when the Web server they are accessing is being updated. If a majority of Web servers in the farm are unable to accept an update, WebSpective automatically rolls back the update to its previous state. It also provides a panic rollback capability to allow an administrator to revert to a previous base version of content after an update has been distributed.

Operations personnel are given the ability to control application configuration and status, add and remove hosts, add and remove Web servers, deactivate and reactivate Web servers for maintenance and updates, and view the status of all site components. All these changes can be applied both to individual Web servers and the entire Web application involving many Web servers across the WAN with full security.

WebSpective is a software solution with components that monitor, manage, and gather data about Web sites. Each component is designed to be nonintrusive in order to minimize the effect of its presence on the site and to prevent site failure should one of the system components develop problems. Principal components are decribed as follows (Figure 7.9) (Mason, 1999).

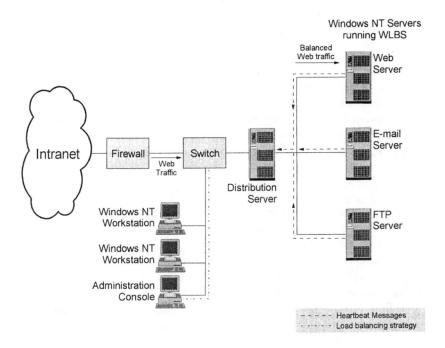

FIGURE 7.9   Architecture of WebSpective.

THE MANAGER

The Manager is the central operational component of the WebSpective architecture. It maintains the master configuration of the entire Web site and records all http "hit" data and all other host data that comes from the Agents to the logging database. It also directs the Agents when to start and stop processes according to user actions and internal rules, updates the Interceptor with dynamic load information for its traffic management decisions, and reacts to alerts from the Agent and the Interceptor.

The Management Console provides administrators with access to the Manager. Through the Management Console, the administrator accesses the Manager to set up application configurations, establish administrative rules, and otherwise take actions on the system. The Management Console graphically depicts the physical and logical layout of the site, as well as the interrelationships and the real-time status and performance level of components and applications in the system. It offers three different views:

1. Management view, which describes configuration and site components
2. Health view, which presents historical and near real-time status of individual site components or of groupings of them in a graphical manner
3. Event view, which decribes all events that have been generated by any site elements

There can be many different Consoles running against the same Manager from different desktops, each independent of the other.

THE AGENTS

Agents run on each Web server. They are responsible for a host and for all Web servers on that host. They report "hit" data and other performance statistics to the Manager, restart failed Web servers, and notify the Manager of Web servers that cannot be restarted. The Agent also processes requests from the Manager that affect the Agent itself and the Web servers it controls.

THE FILTER

The Filter is a shared library that is loaded into the site's Web servers and allows complete control and data gathering about the Web server by WebSpective. The Filter takes advantage of the published APIs for the servers supported by WebSpective. It communicates with the Agent via a private interprocess communication (IPC) channel. The Filter plays a role in administration and presentation, data collection and analysis, crisis management, and traffic management.

The Filter and the Agent may be considered as part of the same component since they only work in conjunction. There is one Agent per host system in the site, while there is one Filter loaded for each Web server. That means, that a host would have only one Agent, but would have one Filter loaded for each Web server running.

THE TRAFFIC MANAGER

The Traffic Manager is the initial portal through which customers gain access to the commerce site. It then redirects all these incoming user requests to the most appropriate Web server, wherever it may be, in order to ensure the appropriate level of service to the customer. The Traffic Manager uses information that it receives from the Manager regarding server loads and administrative changes, including changes in server availability (starts and stops). The Traffic Manager is the first line of security filter against external threats.

THE CONTENT DISTRIBUTOR

The Content Distributor synchronizes and distributes Web site content. The site administrator takes new content and places it on the staging server together with jobs containing rules for taking the content from the staging server and copying it onto the remote hosts. These jobs can have several descriptive attributes. The job can be specified as full replication, indicating that all files should be replicated to all hosts, or as an incremental replication, indicating that only files that are different from what is already on a host should be replicated.

The Distributor then takes the new content from the staging server and passes it to Agents on the designated hosts. In doing so, the Distributor coordinates with the Manager and the Interceptor to direct incoming users around content that is being updated. In this way content updating is performed without a detectable impact on the customer.

The Distributor notifies the administrator of the status and progress of a job. If an error is encountered, the administrator is notified. The Distributor will retry any job that fails unless the administrator instructs otherwise. If there is some sort of problem with the content, after it has been replicated, the Distributor has the capability of rolling back to a previous version of content on the staging server.

THE QuEST

This component provides all the functionality involved in managing and monitoring the service levels of specific applications. QuEST allows the administrator to group URLs as applications, and then monitor their performance, set performance and other statistical thresholds, set alert mechanisms, and gather usage statistics.

The applications or products defined within QuEST start as a collection of URLs for all the pages associated with a particular function on the Web site. Then these URLs are associated with servers, geographic sites, and product profiles or categories. Thresholds and alerts can be set for specific URLs, servers, and products, and for monitoring statistics such as number of hits, number of aborted requests, or response time.

This module pulls this real-time data from the SQL database used by the Manager to store all the data collected by the Agents. The user interface for QuEST is designated for easy alerting and drilldown to identify problems quickly and understand their relationship to the business application performance.

In summary, WebSpective meets most requirements of a mature product to support Web operational management.

7.7.3   BRIGHT TIGER CLUSTERCATS

Bright Tiger ClusterCATS integrates four critical services required for building and controlling highly available, high-performance server infrastructures:

- Monitoring services span content, application and transaction resources, servers, and the network, providing the complete picture of the environment. The monitors report on the status and load of servers, but also on network distances between client and server locations. Content, application, and transaction resource monitors identify the location and availability of dependent resources including databases, applications, files, and other server processes.
- Direction services exploit the intelligence of monitoring services and transparently direct requests to the optimum server. These services direct user requests to the closest, least-loaded server with available CAT resources, providing protection from overloaded servers or downed applications or databases.
- Content distribution services intelligently and automatically place and synchronize Web server content across the SmartCluster for redundancy

and optimal retrieval. Content is added, updated, or deleted between servers by a policy-based system that uses a publisher/subscriber model. Content can be published to or from any server, including staging servers, according to administrator-defined schedules.

• Administrative services manage the tasks of creating, maintaining, and reporting on the SmartCluster. Administrators can add and remove servers from the SmartCluster in real time, configure the availability of server and CAT resources to perform maintenance without interruption of service, and monitor and track overall server and content utilization levels. The administrative services provide automatic notification of any application, database, and server failures and can perform recovery operations on an automated basis.

Integration of these services provides ten capabilities for supporting business-critical Web-based applications:

1. Transparent server fail-over, removal, and addition — Each Bright Tiger ClusterCATS server continuously monitors the availability of other servers in the SmartCluster. In the event of a failure because of a system hardware or Web server software problem, user requests are automatically intercepted and redirected to another server in the SmartCluster across the LAN or WAN. Servers can also be easily removed from the SmartCluster, transparent to users, to upgrade hardware or software, back-up the system, or perform other system administration tasks. User requests are automatically redirected to another server in the SmartCluster. Servers can also be added to the SmartCluster transparently to users. Through a drag-and-drop feature, content on existing servers can be easily replicated to new servers and keep content synchronized. Transparent server fail-over, removal, and addition ensures availability while minimizing administrative requirements.
   Benefits: High availability, manageability, scalability, resource effectivity

2. Transparent server load balancing — Bright Tiger ClusterCATS servers automatically load balance user requests across the LAN or WAN to servers that are "equal" in terms of content, application and transaction resource availability, and network distance to the client. Bright Tiger prevents a server from becoming overloaded by redirecting a percentage of all requests as the server becomes increasingly busy. Bright Tiger's intelligence complements both standard and round-robin DNS configurations. Load balancing minimizes access times for users and optimizes server utilization. Load balancing is beneficial for many applications; however, applications that maintain state on the server in a file or database should not be load balanced at each request. Bright Tiger provides a simple configuration option to ensure that users are not redirected away from a server that is maintaining state.
   Benefits: High performance, scalability, resource effectivity

3. Transparent, network distance–based server selection — Bright Tiger factors network distances between client and server locations into the server selection decision. The selection of the optimum server for any user request is based on the location of the nearest server that can satisfy the request. A server can satisfy the request only if the server is available, is not overloaded, and has the requested content, application, and transaction resources available, including databases, applications, files, and server processes. Minimizing the network distance between client and server locations ensures the fastest possible access times and minimizes WAN bandwidth utilization.

   Benefits: High performance, scalability, resource effectivity

4. Automated, selective content distribution and synchronization — The content distribution service places and synchronizes content across a SmartCluster intelligently and automatically, ensuring availability of the accurate information. Content can include HTML and ASP files, images, scripts, applications, and other entities in the Web service directory. With Bright Tiger's auto-discovery capability, not all content needs to be replicated on all servers in the SmartCluster. Two replication services are supported:

   • Publisher/subscriber
   • Drag and drop

   The publisher/subscriber service is a production-level, policy-based service. Content can be replicated across the LAN or WAN from any publisher location to any subscriber location anywhere in a SmartCluster or from designated staging servers, providing maximum flexibility. A publisher location can be any directory or subdirectory. Bright Tiger's monitoring service automatically discovers content changes (additions, updates, and deletions) and automatically replicates only the changes according to administrator-defined schedules. Synchronization is maintained by placing subscriber locations in a "restricted" state and redirecting requests until replication is completed. Any replication failure to a subscriber is automatically requeued and an alarm generated. Content integrity is maintained through periodic checks of subscriber locations and automatic replication of any missing files.

   The drag-and-drop service provides a simple, convenient mechanism for quickly adding, deleting, or moving content to a new or existing server. Content can be dragged from either a Windows NT Explorer or Bright Tiger ClusterCATS Explorer window and dropped on any destination server or directory locations in ClusterCATS Explorer.

   The content distribution and synchronization services make it extremely easy to replicate content over the LAN or WAN on distributed servers close to users. Local access, typically over a LAN infrastructure, ensures minimum retrieval times. Selective distribution minimizes WAN bandwidth utilization and server hardware requirements.

   Benefits: High performance, manageability, scalability, resource effectivity

5. Location auto-discovery content, application, and transaction resources — Bright Tiger auto-discovers all content in each Web server directory including HTML and ASP files, images, scripts, applications, and other entities. The database monitoring capability of the product also auto-discovers all database replicas on remote server systems. Location auto-discovery eliminates the need for administrators to manually configure and maintain the location of selectively distributed resources.
   Benefits: Manageability, scalability, resource effectivity

6. Content, application, and transaction resource monitoring — Bright Tiger recognizes the trend to "n-tier" Web application architectures to deliver dynamic content and e-commerce applications. The product continuously monitors the availability of all critical content, application, and transaction resources including databases, applications, files, or server processes on Web servers or logically connected remote servers. If any resource becomes unavailable, requests are immediately and transparently redirected to another server with available resources and an alarm is triggered. Local Web server processes can be automatically restarted and, if successful, are reinstated in the SmartCluster to service user requests.
   Benefits: High availability, manageability, scalability, resource effectivity

7. Automated, transparent direction to selectively distributed content, application, and transaction resources — Bright Tiger exploits its monitoring services to automatically and transparently direct user requests to the optimum Web server. The optimum server is that server that satisfies each of the following three requirements:
   • Server has requested content, application, or transaction resources available
   • Server is closest to user in terms of network distance
   • Server is available and not overloaded
   The product can factor all three of these requirements and ensures that users can rapidly and reliably access content, application, and transaction resources while minimizing WAN bandwidth, server, and operational costs.
   Benefits: High availability, high performance, scalability, resource effectivity

8. Distributed operations architecture — The product uses a distributed operations model in which all servers in a SmartCluster are peers in terms of global, cluster-level intelligence. All servers share knowledge of critical cluster availability parameters including content, application, and transaction resource location and availability; server state and load; and client/server network distance. It exploits highly efficient algorithms for SmartCluster monitoring, content distribution, and administrative services. Inter-server communications and replication processes exploit high-performance RPC mechanisms, minimizing network overhead and enabling SmartClusters to currently scale to tens of servers with tens of thousands of URLs.
   Every server in a Bright Tiger SmartCluster can redirect requests to another server if necessary. Once connected to the optimum server, user

requests and responses will flow to and from that server directly, minimizing response times throughout the user session. It does not use a single, centralized server or other network device to redirect or reroute all requests and responses, thus eliminating traffic bottlenecks.

Benefits: High availability, high performance, scalability, resource effectivity

9. Centralized management from any location — Bright Tiger's Cluster-CATS Explorer, the SmartCluster management software, operating in conjunction with an administrative agent on each Bright Tiger server, provides all the required tools for managing one or more SmartClusters from any location — the operations center, hotel, or home. The Explorer easily connects with a SmartCluster to provide the current cluster view. Using a Windows Explorer-like graphical user interface, the tool set supports configuring, viewing, monitoring, and other management tasks for one or more SmartClusters in a single window simultaneously. Centralized management enables easily building scalable, distributed infrastructures and reduces operational costs.

   Benefits: Manageability, scalability, resource effectivity

10. Broad range of SmartCluster configurations — The software tools of the product cost-effectively build and manage a broad range of SmartCluster solutions, satisfying the needs of end users, Web hosters, and integrators. Solutions range from small business Internet hosting applications using two systems with many virtual server-based SmartClusters to enterprise intranet/extranet/Internet applications using a single SmartCluster of distributed servers at many geographic locations. SmartClusters can be comprised of servers with heterogeneous CPU, memory, and disk configurations, optimizing server investments. The use of a common tool set for this broad range of configurations enables easy SmartCluster expansion and minimizes personnel training and support costs.

    Benefits: Manageability, scalability, resource effectivity

Table 7.2 displays how these capabilities are supported by monitoring, request direction, content distribution and synchronization, and administrative services.

Bright Tiger's ten capabilities work together to provide five important benefits:

1. Highly available access: Bright Tiger ensures users can reliably access content, application, and transaction resources.
2. High-performance access: Bright Tiger ensures users can rapidly access content, application, and transaction resources.
3. Manageable infrastructures: Bright Tiger's intelligence and automated capabilities make SmartCluster management extremely easy.
4. Scalable infrastructures: Bright Tiger builds scalable SmartClusters in terms of availability, performance, and manageability.
5. Resource effectivity: Bright Tiger minimizes IS deployment costs and eliminates user opportunity costs.

TABLE 7.2
Support of Capabilities by Services for ClusterCATS

Capabilities	Monitoring	Request Direction	Content Distribution and Synchronization	Administration
Server fail-over	x	x	x	x
Server load balancing	x	x		x
Network distance–based server selection	x	x	x	
Content distribution and synchronization	x	x	x	x
Resource location discovery	x			x
Resource monitoring	x	x		x
Transparent direction to distributed resources	x	x		x
Distributed operations architecture	x	x	x	x
Centralized management	x			x
Broad range of solutions	x	x	x	x

7.7.4 WINDOWS NT LOAD BALANCING SERVICE (WLBS)

Web servers are stressed by millions of requests by visitors. Load balancing and distribution is thus gaining importance. Microsoft offers the Windows NT Load Balancing Service (WLBS) to ease loads on servers. The software supporting WLBS distributes requests across as many as 32 NT machines. And, as the first software load balancer tightly integrated with NT, it avoids the single point of failure created by hardware-based balancers. It goes beyond the conventional round-robin schemes typically used by these products, allowing corporate networkers to assign traffic to specific machines in a cluster. On the downside, WLBS cannot load balance across different geographic regions. Its command line interface, instead of a graphical user interface, is criticized by users.

WLBS works with the Windows NT 4.0 (or higher) Enterprise Edition. The 1.5-MB package is loaded onto each server in a cluster; its built-in algorithm distributes TCP/IP traffic across various NT machines — Web, e-mail, and FTP servers, for example — balancing the load according to percentages assigned by the network manager. Because the algorithm sits between the TCP/IP stack and the NDIS (network device interface specification) driver on the server's adapter, it runs beneath the application protocols. That means it introduces little overhead and keeps load balancing operations clipping along (Bruno, 1999).

FIGURE 7.10 Windows NT load balancing service.

WLBS is elegant in its simplicity. Network managers assign each Web server a percentage of the load they want distributed to that machine. All incoming packets arrive at a designated WLBS server, which then passes off processes to others in the cluster. Typically, faster boxes will service a higher percentage of traffic than slower ones. Clustered servers communicate via 1.5-KB "heartbeat" messages that indicate if they are up and running and accepting traffic. The frequency of heartbeat messages can be set by the NT server console. If a server breaks down, the rest of the servers in the cluster divide the traffic that would have gone to the broken server.

Competitive products may offer attributes beyond the functionality of WLBS, such as the inclusion of database servers and load balancing across different geographic domains. Usually, they offer a graphical user interface that simplifies deployment and operations. However, competitive products are not integrated with NT.

Other product examples for Web server load balancers include:

Alteon 180, 708, and 714	Alteon Websystems, Inc.
CS-100 and CS-800	Arrowpoint Communications, Inc.
Distributed Director 2501 and 4700	Cisco Systems, Inc.
Equalizer 250, 350, and 450	Cayote Point Systems
3 DNS appliance	F5 Labs, Inc.
Serveriron	Foundry Networks, Inc.
HyperFlow 2	HolonTech, Inc.
HydraHydra 100	Hydraweb Technologies, Inc.
e-Network Dispatcher	IBM
Ipivot Multisite Director 9000	Ipivot, Inc.

7.8 SUMMARY

The basics of server management may be easily applied to Web servers. The principal functions of server monitoring, workload management, performance management, capacity management, and sizing resources are the same or at least very similar to generic server management. Products supporting Unix and NT server management can be easily adapted to Web server management. The new performance challenge has its roots in the unusual workload profiles driven by unusual visitor behavior. To reduce the stress on single Web servers, Web servers can be operated in various forms of load sharing, called Web server farms. Load sharing may show different forms, such as full replication of content, distribution of content by subjects or randomly, or distribution by traffic profiles or by Web server power between Web servers of the farm. A few emerging products address content-smart Web server performance analysis and optimization.

8 Load Balancing and Optimal Distribution

CONTENTS

8.1   INTRODUCTION

To help IT managers track IP performance and optimize bandwidth usage across WANs, several new vendors are offering hardware- and software-based load balancing products. Load balancers typically reside at the edges of corporate networks and determine traffic priorities. They apply a policy that defines different traffic types and determine what happens to each. A very simple policy may call for priorities for a specific sender. Other criteria may be TCP port numbers, URLs, and DNS (domain name service). Traffic shaping may be supported by queuing or via TCP rate control. Products are available for both categories.

Optimization is accomplished by controlling enterprise traffic flows at the boundary between the LANs and the WAN. Because these products give priority to traffic according to application type or even individual user, they allow IT managers to take the first steps toward policy-based quality of service (QoS) in their networks. These products are a logical evolution from the passive probes, which gave users a certain level of visibility for fault operations monitoring but no actual control over traffic. These products go further and can manipulate traffic. IT managers expect that this new class of traffic-shaping tools will ease the contention for bandwidth without forcing them to purchase more and larger physical transmission lines.

This chapter introduces a couple of innovative solutions provided by start-ups and known flow-control companies.

8.2   THE NEED FOR BANDWIDTH, SERVICE QUALITY, AND GRANULARITY

Bandwidth management is rapidly becoming a must for Internet service providers as well as corporations running their own global intranets. The reasons for bandwidth management are the following:

- The move to Internet/intranet-based business
- The need for guaranteed bandwidth
- Service level agreements
- The need for granularity

8.2.1   THE MOVE TO INTERNET/INTRANET-BASED BUSINESS

Corporate networks are rapidly evolving from a classic client/server paradigm toward an intranet-based model, based on information sharing and Web navigation. Analysts predict that by the year 2000 there will be more than 3 million private intranet sites, compared to approximately 650,000 Internet sites. The result is the demand for significantly more bandwidth. Adding more channels and more bandwidth to each channel will not guarantee availability and performance where it is needed most. An intranet-based model implies the following:

- Changing patterns of network use and unpredictable demands for bandwidth — Global users access the network 24 hours a day, 7 days a week. As information appears and disappears on Web sites, access patterns change and saturation moves around the network.
- Demand for increased amounts of bandwidth — People may stay on the link for extended periods of time and download large amounts of data.
- Demand for guaranteed quality of service in terms of bandwidth and minimum delay — Emerging Internet applications are both bandwidth intensive and time sensitive, often requiring support for voice, video, and multimedia applications across the network infrastructure.
- Lack of control by IT staff — Workgroups and departments generally create their Web sites without IT approval, generating increased traffic without necessarily having the infrastructure to handle it. This often results in excessive traffic at the fringes of the network where Web sites are situated, generating traffic precisely where there is the least provision.
- A change in user attitude — Users expect instant access to information without delays or restrictions, especially if that information is critical to their work.

8.2.2   THE NEED FOR GUARANTEED BANDWIDTH

Current networking technology has two major limitations:

- The bandwidth available on a link at any given moment cannot be predicted in terms of quantity or quality — Bandwidth management is needed to allow applications that require a specific QoS, in terms of bandwidth and delay (such as desktop video conferencing), to reserve the bandwidth quality of service they need.
- It is difficult to control which applications or users get a share of the available bandwidth — In some circumstances, an application or a user can take control of all the available bandwidth, preventing other applications or users from using the network. To solve this problem, the user can either add extra capacity at additional costs, resulting in an overprovisioned network that still does not guarantee equal access, or the user can introduce bandwidth allocation.

8.2.3 THE NEED FOR SERVICE LEVEL AGREEMENTS

Virtual private networks (VPNs) are a popular value-added Internet service that corporations are increasingly moving toward. Enterprise customers seeking a VPN provider are more likely to sign with an ISP who can offer a contractual service level agreement — one that guarantees quality of service.

Although service level agreements (SLAs) cannot guarantee end-to-end service across the public Internet, they can be implemented for transport over a single-vendor network or for Internet server hosting. In these areas, an SLA is an important differentiator for an ISP.

Generally, the customer subscribes to a particular class of service and signs an SLA accordingly. Packet throughput is monitored as part of the agreement. Value-added services are expected to grow significantly over the next five years. ISPs who want to get a piece of this additional business, clearly need to implement bandwidth management to meet SLAs that guarantee QoS to customers. Only efficient bandwidth management can enable them to tune network behavior so that customers receive the QoS for which they are charged.

The new network service paradigm is a service-driven network. This is a responsive, reliable, modular infrastructure, based on the latest generation of management technology and built on dynamic, flexible management services. To respond to today's business needs, ISPs and large enterprises must deploy the service-driven network. It delivers innovative services, such as unified roaming, push browsers, multicast, online shopping, etc., to customers faster and at a lower cost than ever before.

8.2.4 THE NEED FOR GRANULARITY

Bandwidth allocation based simply on filtering by protocol is not sufficient to meet bandwidth management needs. One of the key issues in this area is the extensive and increasing use of HTML/HTTP systems for OLTP. Within the next few years, the volume of HTTP-based OLTP traffic is expected to exceed the volume of traditional OLTP traffic. A fine level of granularity is needed for bandwidth management to take into account more than just the protocol when assessing the relative importance of network traffic. Bandwidth management must base allocation not only on protocol type, but also on the application and users involved.

8.3 ISSUES OF DEPLOYING LOAD BALANCING PRODUCTS

Load balancing helps managers use resources more effectively. Concurrently, the end user response time may also be stabilized and improved. This is an emerging area with a number of innovative products that are either hardware or software based. There are even a few that implement load balancing functions in both hardware and software. The hardware solution is faster, but software offers more flexibility if changes are required.

The functionality of a load balancer can be deployed in a stand-alone device or embedded into existing networking components, such as routers, switches, and firewalls. The stand-alone solution offers broad functionality without impacting any other routing, switching, or firewall functions. However, it will add components into the network that must be managed. It may add another vendor that should be managed as well. The embedded solution is just the opposite; easier management at a price of conflicting functions with its host.

Load balancers are successful only when policy profiles can be implemented and used. Policy profiles are most likely based on supporting various transmission priorities. Priorities may be set by applications, by users, or by a combination of both. The technology of the solution may differ from case to case and product to product, but most frequently the TCP flow is intercepted.

Load balancers are expected to support a number of services, such as quality control, resource management, flow control, link management, and actually load balancing. Advanced products support all these services in dependency of page content. It requires more work to gather the necessary information about content, but it offers better services for high-priority content.

Its functions in a narrower sense include traffic shaping, load balancing, monitoring, and baselining. Baselining means to find the optimal operational conditions for a certain environment. It may be expressed by a few parameters, such as resource utilization, availability, and response time. Load balancers should monitor these few metrics and act on them. Traffic shaping and load balancing help the system get back to "normal" by splitting traffic, redirecting traffic to replicated servers, delaying payload transport, etc.

One of the most important questions is how load balancing performs when data volumes increase. Volume increase can be caused by offering more pages on more Web servers, more visitors, longer visits, or extensive use of page links. In any case, collection and processing capabilities must be estimated prior to deciding on procedures and products.

Load balancing products can be managed by SNMP or Wbem agents. They are handled by managers just as are other kinds of managed objects. Table 8.1 summarizes selection criteria for load balancers.

Documentation may have various forms. For immediate answers, an integrated online manual would be very helpful. Paper-based manuals are still useful for detailed answers and analysis. This role, however, will soon be taken over by Web-based documentation systems. In critical cases, a hotline can help with operational problems.

Managing load balancers out of a management platform offers integration at the management applications level. Baselining and monitoring may even be supported by other applications. In the case of management intranets, universal browsers may be used to view, extract, process, and distribute management information. The only prerequisite is that Wbem agents have been implemented and that CIM is supported for information exchange.

8.4   CONTENT-DRIVEN POLICIES

Although Web server performance improvements are part of the performance optimization solution, they must be accompanied by improvements in network and content management technology to have a true impact on WWW scaling and performance. Specifically, developments in the following three areas are critically important:

TABLE 8.1
Selection Criteria for Load Balancers

Solution is hardware or software based
   Hardware
   Software
   Combination
Solution is stand-alone or embedded
   Stand-alone traffic shaper
   Embedded into router, switch, or firewall
Support of policy profiles
Support of transmission priorities
Technology of solution
   TCP flow intercepts
   Others
Content dependency
   Quality of service
   Resource management
   Flow admission control
   Link management
   Load balancing
Major scopes of functions
   Traffic shaping
   Load balancing
   Monitoring
   Baselining
Scalability
Manageability of the product if stand-alone
   SNMP agent
   Wbem agent
Integration with management platforms
Documentation and support
   Integrated online manual
   Paper manual
   Help desk
Vendor
   Number of clients
   Founded when
   What other products are offered
   Support (e.g., on-site, hotline)
   Maintenance contracts
   Financial strength
   Keeps current with servers software releases

- Content distribution and replication — By pushing content closer to the access points where users are located, managers can reduce backbone bandwidth requirements and response time to the user. Content can be proactively replicated in the network under operator control or dynamically

replicated by network elements. Caching servers are an example of network elements that can facilitate the dynamic replication of content. Other
devices and models are likely to emerge over time.

- Content request distribution — When multiple instances of content exist in
  a network, the network elements must cooperate to direct a content request
  to the "best fit" server at any moment in time. This requires an increasing
  level of "content intelligence" in the network elements themselves.
- Content-driven Web farm resource measurement — A server or cache in
  a server farm ultimately services a specific content request. Local server,
  switching, and uplink bandwidth are precious resources that need to be
  carefully managed to provide appropriate service levels for Web traffic.

8.5 LOAD BALANCING AND OPTIMIZATION TOOLS

This section will introduce different products. Some of them are hardware based,
some of them are software based, and some use a combination of both.

8.5.1 AC200 AND AC300 FROM ALLOT COMMUNICATIONS

Allot Communications delivers a complete, active traffic management system that
maximizes the efficiency of network and server resources. The Allot system allows
network managers to set policies that categorize, monitor, and prioritize traffic
travelling between high-speed LANs and slower speed WANs.

This product family gives users the power to intelligently shape network bandwidth and balance server traffic based on the needs and priorities of the corporation.
During peak periods, the Allot system will assign less bandwidth to lower priority
applications to deliver more bandwidth and server resources to mission-critical and
time-sensitive applications.

The Allot product family consists of three products:

- The AC200 is a hardware and software solution that is optimized for small
  to medium size companies. It supports WAN speeds up to 1.54 Mbps (T1)
  and is optimized for networks of up to 1000 users.
- The AC300 has added capabilities, in addition to those of the AC200, for
  supporting larger size network configurations. The AC300 supports speeds
  up to 45 Mbps (T3).
- The load-balancing module is a software add-on package that runs with
  either AC200 or AC300 products. The module provides capabilities for
  balancing the applications between multiple servers in an Internet or WAN
  server farm.

8.5.1.1 Architecture

The Allot architecture is shown in Figure 8.1. This diagram illustrates two applications of the Allot system. Placed between the LAN and the WAN, the Allot system
continuously monitors, prioritizes, and shapes network bandwidth based on user-
defined policies. Placed in front of corporate Internet servers, it integrates server
load balancing and traffic shaping capabilities.

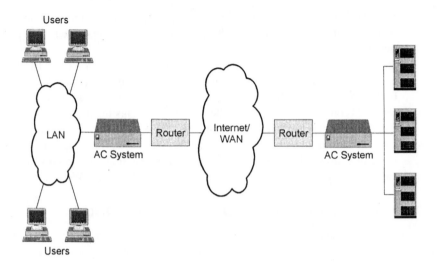

FIGURE 8.1   The AC architecture from Allot.

The Allot system analyzes incoming traffic and applies user-defined policies that are applicable to that particular type of traffic. Users can group and define traffic according to high-level, easy-to-understand concepts. Traffic can be logically classified into categories such as

- Mission-critical accounting application
- Timing-sensitive video feed
- Low-priority Web traffic

8.5.1.2   Virtual Channels

These categories or virtual channels are automatically translated to physical control of network resources. Virtual channels are used to define end-to-end traffic characteristics based on:

- IP source and destination address
- TCP and UDP port number
- Time of day
- Concept inspection — Virtual channels can also be defined using higher application level classification, such as
    HTTP URL, including specific directories
    Web content type, including specific Web pages or categories of Web traffic

Each virtual channel is assigned specific classes and quality of service:

- Prioritized bandwidth delivers levels of service based on a connection's import demand for traffic relative to other channels. Users can choose from the different levels of priority for any virtual channel. During peak traffic periods, the Allot system will slow down lower priority applications and deliver more bandwidth to those with higher priority. By assigning priority based on what users expect, it ensures that everyone receive predictable service.

- Guaranteed bandwidth allows the assignment of a guaranteed amount of bandwidth to specific virtual channels. Guaranteed bandwidth includes:

    Minimum or maximum bandwidth for a virtual channel
    Minimum bandwidth for a connection
    Constant traffic rate (min equals max) for a connection

    By assigning guaranteed bandwidth, it ensures that timing-critical applications will receive a constant level of service during both peak and non-peak traffic periods.

Advanced policies can be defined for specific virtual channels. Channels can be defined to:

- Reject a connection
- Deny a connection
- Limit connections

8.5.1.3 Traffic Monitoring and Reporting

The Allot system's Java-based management and reporting tools allow users to understand their network characteristics and gives them the ability to create intelligent policies that result in efficient network performance and behavior.

The Allot system allows users to create a complete system-wide approach to managing the network. It will constantly monitor network bandwidth and server resources and report the status in a concise, easy-to-understand format that can be accessed from any remote Web browser. Users can define policies based on precise monitoring profiles. As network and server performance changes over time, the user can continue to monitor resources and refine policies to create an optimized network system.

Statistics collection includes:

- Top connection originators
- Top connection destinations
- Per virtual channel traffic distribution
- Per protocol traffic distribution

Example application environments for the Allot product family include the following:

- Internet service providers can use the Allot system to deliver higher levels of service to priority customers.
- Web hosting services will find the Allot system ideal for balancing and prioritizing traffic to multiple customers sharing common server resources.

- Corporate network managers can control traffic flow from Web-based customers and remote offices to centralized corporate networks and services.
- Remote or small offices with slow WAN or Internet connections need to use the Allot system to prioritize user access to all remote resources.

8.5.1.4 Server Load Balancing Module

This module, in conjunction with the traffic shaping capabilities of AC200 and AC300, allows users to define single policies that control both the prioritization of network bandwidth and the distribution of clients to corporate servers. The Allot system continuously adjusts the flow of applications through the network by distribution of those applications to servers, resulting in efficient, optimized traffic through the network. The system will balance traffic according to the same virtual channel policies used to prioritize network traffic including addresses, TCP port, and Web content type. With the Allot load balancing module, server traffic does not need to be mirrored between servers. Applications can be distributed between the servers according to their individual capabilities. The module can perform network address translation functions, allowing a single IP address to represent multiple servers. Traffic can be balanced between servers using algorithms such as:

- Round-robin, whereby each server will be treated with equal priority
- Weighted round-robin, whereby each server can be given an individual priority weight based on its ability to deliver specific applications
- Maintenance rerouting, by rerouting traffic to another server when the server becomes unavailable

Allot Communications' products maximize the efficiency of network and server resources. They provide a structured, policy-based approach to ensuring the quality and availability of network services and resources.

8.5.2 FLOODGATE-1 FROM CHECK POINT SOFTWARE TECHNOLOGY, INC.

FloodGate-1 solves the network congestion problem by delivering a policy-based, enterprise-wide, bandwidth management solution. By dynamically managing the overall mix of communication traffic, it minimizes congestion on oversubscribed Internet and intranet links, allowing network administrators to allocate limited bandwidth resources based on criticality or merit. In addition to increasing network efficiency and improving user experience, it minimizes the addition of costly bandwidth. Bandwidth management policies are defined by the graphical user interface. The management policy consists of traffic rules that define bandwidth privileges for user-defined traffic classes. Designed as a client/server application, FloodGate-1 allows a single enterprise policy to be defined and automatically distributed to multiple FloodGate-1 modules. Figure 8.2 shows an example for policy management.

FloodGate-1 Policy Manager

Bandwidth Policy

Rule Name		Source	Destination	Service	Control	Installation
Inbound_Web	Weight 50	Any	Web_Servers	HTTP	Weight = 50	All
Outbound_Web	Weight 25	Loca_Net	Any	HTTP	Weight = 25	All
Point Cast	Weight 1	Any	Any	Pointcast	Weight = 1 Limit = 10,000	All
Video_Conference	Weight 30	Remote_Office	Management_Group	NetMeeting	Weight = 30 Guarantee = 30,000	All
Sales_VP-Exception	Weight 40	Sales_VP	Sales_Server	FTP	Weight = 40 Exception = true	All
Field_Sales	Weight 10	Field_Sales	Sales_Server	Any	Weight = 10 Guarantee = 5,000 Limit = 20,000	All
Default	Weight 10				Weight = 10	All

FIGURE 8.2 Policy management control table from Check Point.

It uniquely delivers total bandwidth management by controlling the bandwidth usage of entire classes of traffic, not just individual connections. Organizations can now allocate valuable bandwidth resources among multiple classes of traffic, such as critical applications or important groups of users. The product provides powerful control criteria to actively manage bandwidth consumption by traffic class. In addition to basic bandwidth guarantees and limits, all network traffic can be controlled using weighted priorities that allocate bandwidth based on relative merit or importance. Unlike absolute priorities, weighted prioritization ensures that all traffic is delivered and not starved of bandwidth.

A real-time traffic monitor is integrated with the product to analyze network congestion and monitor bandwidth allocation. Comprehensive information is provided on all inbound and outbound network traffic to define custom bandwidth management policies and to monitor the effect of existing policies. This feedback mechanism allows precise adjustment of bandwidth policies to meet specific management requirements. FloodGate-1 is a stand-alone bandwidth management solution that can be used with any Internet firewall. When integrated with FireWall-1, however, organizations can support their existing VPN deployments and provide flexible address translation.

Leveraging Check Point's patented stateful inspection technology and its innovative intelligent queuing engine, FloodGate-1 precisely controls the bidirectional flow of all IP-based traffic. Total bandwidth management is achieved by controlling the bandwidth usage for entire classes of traffic. It also provides real-time traffic monitoring capabilities to diagnose the source of network congestions. Bandwidth can be allocated for mission-critical and high-priority applications, and the burst-and-delay effect inherent in most Internet traffic can be eliminated.

By dynamically controlling the mix of traffic not just on a per-connection basis, but at an aggregate level, FloodGate-1 delivers total bandwidth management. It classifies traffic by:

- Internet service (HTTP, FTP, Telnet, etc.)
- Source
- Destination
- Group of users
- Groups of Internet services
- Internet resource (e.g., specific URL designators)
- Traffic direction — inbound or outbound

Once traffic is accurately classified, FloodGate-1 applies one or more control criteria to dynamically manage bandwidth allocation. Primary control criteria include weighted priorities, limits, guarantees, and guaranteed allotments. These control criteria can be used alone or in concert to define flexible policies to meet specific needs. Weights allocate available bandwidth based on relative merit or importance.

Weighted priorities enable the network administrator to define the basis upon which different classes of traffic compete for available bandwidth. An unlimited number of weighted priorities can be defined to intelligently divide finite bandwidth among multiple classes of traffic.

Weighted priorities provide an intuitive method of defining bandwidth policies. As an example, outgoing Web traffic can be deemed twice as important as incoming FTP traffic. When bandwidth resources are oversubscribed, FloodGate-1 will ensure that the ratio of outgoing Web traffic to incoming FTP traffic is accurately maintained at 2:1.

Guarantees ensure minimum bandwidth for user-defined classes of traffic which in turn ensures consistent network performance and smoothing the burst-and-delay effect common in IP-based traffic. This feature improves the quality of service for streaming applications such as video conferencing. The implementation of guarantees, unlike basic bandwidth reservations, allows unused bandwidth resources to be dynamically allocated to other classes of traffic for optimum network efficiency.

Guaranteed allotments set the total amount of bandwidth that is allocated to meet all bandwidth guarantees. A network administrator can, for instance, commit 25% of a T1 line to meet the bandwidth guarantees defined in all traffic rules. Limits set bandwidth restrictions for noncritical network services or for applications that do not require large amounts of bandwidth. A typical implementation would be to limit the allocation for bandwidth-intensive "push" technologies. Requests that exceed defined limits compete for additional bandwidth based on user-assigned weights.

FloodGate-1 leverages a flexible three-tier client/server architecture, allowing a single bandwidth management policy to be distributed throughout an enterprise network. This solution consists of three components, all of which can run on a single platform or be distributed across multiple workstations/servers for maximum flexibility. The graphical user interface has a native Windows look and feel and contains

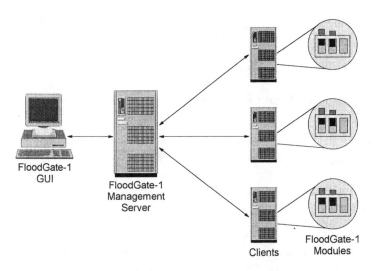

FIGURE 8.3   Tiered architecture of FloodGate-1.

the policy editor, which is used to create and modify the bandwidth management policy and define the network objects. As a Java application, this GUI can run on a variety of platforms. The management server controls and distributes the bandwidth management policy to all FloodGate-1 modules. It also stores all logging and monitoring data. The modules of the product enforce the bandwidth management policy at network access points. It includes the INSPECT virtual machine and the IQ engine.

Figure 8.3 shows the three-tiered architecture. This flexible architecture is scalable and enables a single management policy to be defined and automatically distributed to multiple FloodGate-1 modules for policy enforcement. All modules, local or remote, can be centrally managed from a single management console.

The most important features can be summarized as follows:

- Allocates bandwidth resources with a single, centrally managed enterprise-wide policy
- Prioritizes bidirectional network traffic using weighted priorities, limits, and guarantees
- Supports virtual private networks and network address translation
- Scales from low-speed dial-up lines to high-speed network connections
- Monitors and analyzes bandwidth consumption by user, application, connection, and direction
- Java application with same look and feel across all platforms

The benefits of the product include the following:

- Diagnoses and alleviates traffic congestion on Internet and intranet links
- Ensures reliability and optimum performance of mission-critical business applications

- Increases network efficiency by reducing the retransmission of data
- Minimizes the need to add costly bandwidth
- Integrates bandwidth management with security management policies for secure enterprise connectivity

8.5.3 FORT KNOX POLICY ROUTER FROM INTERNET DEVICES, INC.

This product family is a software-extendible modular solution that consolidates full-featured virtual private networking and enterprise-level firewall security, along with other key integrated services such as URL filtering, Web caching, real-time monitoring, and e-mail. Fort Knox is a turnkey hardware/software offering that works in conjunction with, and adds value to, today's leading network routers including those from Cisco, 3Com, and Bay Networks. There are three versions of the product:

- Fort Knox 1000, which supports up to 512 Kbps connections
- Policy Router 3000, for sites with T1/E1 connections
- Policy Router 5000, for sites with multiple T1/E1 connections

The Fort Knox family of products combines the following services in one rack-mountable device:

- Standards-based virtual private networks
- Enterprise-level firewall security
- Application- and device-level user access control
- 10 Mbps or 100 Mbps connectivity
- Network address translation
- URL blocking with CyberNOT™
- E-mail server
- DNS caching
- Web caching
- Real-time network monitoring
- Internet usage reports
- Extensive real-time network diagnostics
- SQL database for generating custom reports
- Extensive policy management capabilities
- In-band and out-band Web-based management tools

The product can be easily installed and configured. Internet Devices has developed a unique single-IP address configuration, meaning the product can be added to an existing network without requiring address changes on the Internet router, personal computers (PCs), or other network-connected devices. The complexity of installation users normally experience with traditional, single-function solutions is eliminated.

8.5.4   GUIDEPOST FROM NETGUARD, INC.

Guidepost is a powerful tool that enables the system administrator to easily and efficiently manage and control access to Internet, intranet, or extranet connections. The nature of TCP/IP communication creates a "first come, first served" environment that allocates bandwidth on a fluctuating and unpredictable basis without taking into account the real priorities of the organization. Guidepost regulates and controls bandwidth allocation, letting the user set minimum and maximum levels of service. The user can effectively allocate available bandwidth according to predefined priority categories that form part of an overall network strategy. This strategy can be modified whenever necessary or even create a library of strategies that lets the user allocate Internet and other service resources to the various users and departments in the organization according to changing network needs. Guidepost contributes to good system administration by providing the network manager with a powerful tool for monitoring network resources, enabling real-time troubleshooting and control.

The Guidepost system consists of two functional modules — a manager and an agent. The agent module inspects and controls network traffic. The manager controls the agents, installing and modifying the network bandwidth allocation strategy when needed. The agent is installed between the Internet and the local-controlled network. It inspects every packet passing through it, allocating the appropriate bandwidth to all traffic between the network and the Internet according to a predefined bandwidth control strategy. The manager module may be installed on a suitable host either at a convenient network site or anywhere in the Internet. Authorized to do so, one network manager can control one or several distributed networks. The system data flow is shown in Figure 8.4.

This product provides the following powerful bandwidth control and network traffic monitoring features:

The Guidepost strategy — This defines the network bandwidth control policy, determining which users or groups of users will be allocated guaranteed, prioritized, or standby bandwidth allocation and setting minimum and maximum allocations for each category and group.

The Guidepost tree — This allows the customer to create collections of users or services and represent them as objects. These objects can then be assembled into convenient groups and subgroups, permitting them to be handled either as individual items or as part of increasingly larger, more complex objects. The tree provides the network administrator with the capability to immediately and efficiently control a group of users, a domain, or a department with a few mouse clicks.

Real-time monitoring — This allows network administrators to monitor four simultaneous graphical or numerical network traffic views for each agent, enabling the customer to analyze the unique patterns of the networks and solve specific bandwidth allocation problems.

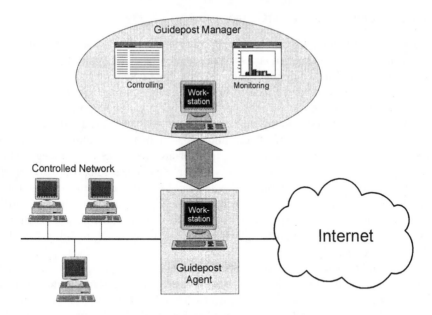

FIGURE 8.4   Guidepost system data flow.

Organizations that deploy Guidepost will benefit from:

- Predictability — Standard Internet QoS constantly fluctuates depending on enterprise demand and competition. Simply installing Guidepost, without any configuration, results in an immediate stabilizing and equalizing of bandwidth resources for all users.
- Guaranteed bandwidth — Guidepost can allocate bandwidth to mission-critical servers, subnetworks, or users at guaranteed levels, providing them a fixed minimum of bandwidth resources.
- Priority — Guidepost can provide differential QoS by creating priority levels. A user with priority 1 will receive three times the bandwidth than another user with priority 3.
- Standby — Guidepost can move noncritical traffic away from peak hours by defining a service as standby.
- Limit — Guidepost can be used to limit allocation of bandwidth to a user group or server, thereby freeing resources for critical tasks.

8.5.5   WISEWAN PRODUCT FAMILY FROM NETREALITY

With the growing importance of corporate intranets and links to the Internet backbone, many corporations are experiencing congestion in core business transactions such as sales, logistics, and finances. Bandwidth-hungry applications such as video conferencing and electronic commerce often must compete with other applications for WAN bandwidth. Corporations see WAN bandwidth as a strategic communications resource, demanding tools that let them maximize WAN access rates, optimize quality of service, and regain control over the usage of their precious bandwidth so that business-critical applications run unhindered. ISPs and carriers are seeking ways to add competitive differentiators, create new revenue streams, and cut their own operating costs.

WiseWan sets a new standard as an open, vendor-independent WAN platform for internetworking services. Combining power with flexibility, WiseWan is the first platform to support the integration of multiple applications, an absolute necessity in a fast-changing WAN environment. Current features include real-time monitoring and analysis, adaptive traffic shaping, and global WAN management, with many more to come, both from NetReality and its technology partners. It delivers:

- A modular, multiservice, scalable platform
- A true adaptive shaping based on real-time monitoring
- A solution designed for the WAN that sits on the WAN

WiseWan provides control of WAN congestion, effectively allocates scarce bandwidth, and optimizes line utilization at significant operational cost savings. WanShaper provides intelligent, adaptive bandwidth allocation at the WAN circuit level in response to real-time monitoring of ever-changing bandwidth availability. Designed to shape traffic directly on the WAN, WanShaper employs an adaptive circuit-based shaping (ACS) algorithm to optimize bandwidth utilization on the fly, significantly cutting internetwork expenditures.

WanShaper offers adaptive traffic shaping with real-time on-WAN monitoring, a distinct advantage over every other bandwidth management system. Because it is physically deployed on the WAN, WanShaper can adapt bandwidth allocation according to the actual, constantly changing traffic status on every single circuit. This is the only way to ensure optimal use of bandwidth at all times, so that quality of service is guaranteed for business-critical applications. An individual graphical interface allows the network manager to set priority levels for each user and application. Corporate users can use this facility to ensure that business-critical applications or executive video conferences have maximal bandwidth. ISPs can use this facility to define and implement differing classes of service for their subscribers, creating competitive differentiation and new revenue streams. Figure 8.5 gives an example of application classification.

WiseWan simplifies and automates routine network administration, providing single-point visibility of global WAN traffic, instantly alerting the user if irregularities or service level violations occur. So network managers can concentrate on the big picture, ready for anything, and delivering the reliable QoS end users demand. WanXplorer pinpoints high-volume users and applications, records their network impacts, and suggests cost-saving measures that can reduce operating costs. Constantly communicating with WanSentry monitoring probes, it analyzes WAN traffic at the circuit level in real time. By tracking bandwidth usage and validating CIR (Committed Information Rate) levels, network managers can detect service level breaches and claim appropriate credits from providers.

WanShaper Tree

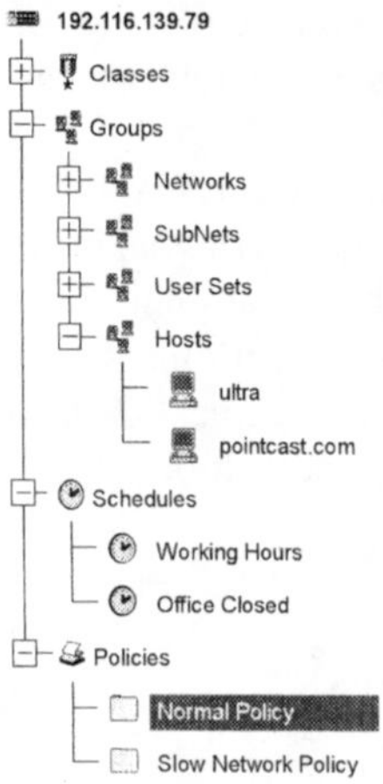

192.116.139.79
- Classes
- Groups
  - Networks
  - SubNets
  - User Sets
  - Hosts
    - ultra
    - pointcast.com
- Schedules
  - Working Hours
  - Office Closed
- Policies
  - Normal Policy
  - Slow Network Policy

Class	Direction	Schedule	Local Group	Remote Group	Priority	Bandwidth Limit	Guarantee Limit	Drop Limit
Business Applications	Outbound	Working Hours	ANY	Corporate Servers	Pass through			20 %
Business Applications	Both	Office Closed	ANY	ANY	Block			0 %
SQL Protocols	Outbound	Working Hours	ANY	SQL Servers	Pass through			20 %
SQL Protocols	Outbound	Office Closed	ANY	ANY	Block			0 %
Video Services	Both	Always	Top Mgmt	Top Mgmt	Pass through		50 K	15 %
Video Services	Both	Working Hours	ANY	ANY	Block			0 %
Voice	Both	Working Hours	Domestic	International	Pass through		Best	
Voice	Both	Working Hours	Domestic	Domestic	High		Normal	
Mail	Both	Always	ANY	ANY	High			20 %
HTTP	Inbound	Always	WEB Servers	ANY	High			20 %
HTTP	Outbound	Working Hours	ANY	ANY	Medium-High			30 %
PointCast	Outbound	Working Hours	ANY	pointcast.com	Low	2 K		
FTP Protocols	Outbound	Working Hours	ANY	ANY	Low			
FTP Protocols	Inbound	Always	FTP Servers	124.138.255.255	Medium-High			40 %
ANY	Both	Always	CIO	ANY	Pass through		100 K	0 %

FIGURE 8.5   WanShaper application classification.

Principal features include:

- One-point global management across multiple WAN links
- Real-time event monitoring, which leads the user directly to the sources of a problem
- Monitoring at a glance, which ensures optimal WAN performance
- Grouping mechanism that lets users manage multiple resources according to preferences
- Color-coded indicators that identify the real-time status of every resource
- Intuitive zoom-in analysis flow supported by a drill-down option between reports
- Real-time troubleshooting, which allows the user to drill down to the circuit/conversation level
- Extensive SQL database, which provides in-depth, customizable reports
- Automatic prediction that allows users to realize trends and respond before they impact availability
- Cost accounting for tracking network usage and SLA validation
- Optimization of current resources, virtual circuits, server placement, and CIR levels
- Web-based technology for remote management from any browser

WanXplorer offers powerful analysis and reporting tools for managing every aspect of WANs. Routine management and analysis functions are performed automatically, leaving network managers free to focus on strategic issues and exceptional events. A color-coded graphical interface ensures an at-a-glance view of WAN health

WanXplorer DLCI Performance Report

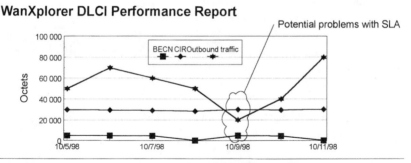

WanXplorer Application Protocol Distribution

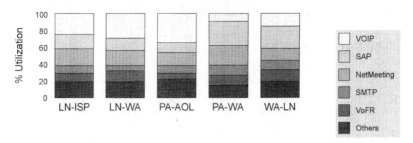

FIGURE 8.6   Reporting examples with WanXplorer.

status, leading the user directly to the source of a problem. An intuitive zoom-in capability offers analysis from lines to circuit level. Reporting examples are shown in Figure 8.6.

8.5.6   NETSCREEN-10 AND NETSCREEN-100 FROM NETSCREEN TECHNOLOGIES, INC.

These products are first of all for the network security market. They combine firewall, virtual private network, and traffic management functionality on a single, dedicated platform. Serving as both a security and traffic control solution, NetScreen will act as a bridge between networked hosts and routers in network environments. NetScreen is pioneering a new approach to network management that will accommodate both today's and tomorrow's needs for high-performance, full-functionality security solutions.

8.5.6.1   NetScreen-10

Its patent-pending system architecture and custom-designed ASIC provide the most secure solution for critical enterprise data.

Firewall attributes include:

- Network address translation (NAT) hiding inside IP addressing
- Dynamic filter, to protect network services
- User authentication, to allow authorized access only

Its design supports wireline performance for critical VPN functions and sets the standard for VPN solutions that include:

- IPSec-compatible interoperability with other network devices
- IKE key management, which secures the exchange of keys
- DES and triple DES, which offer the highest possible level of encryption

Traffic shaping enables network administrators to monitor, analyze, and allocate bandwidth to support network traffic. It offers:

- Traffic bandwidth and prioritization management to manage traffic in accordance with user expectations
- Virtual IP maps and protection for internal servers

NetScreen-10 provides real-time monitoring and logging of network traffic, offering:

- Real-time logging and alarming, ensuring network status monitoring
- E-mail alerts to notify administrators of system and network events

NetScreen-10 can be configured remotely using universal browsers or locally with a Windows 95/NT utility, SNMP manager, or command line interface.

8.5.6.2 NetScreen-100

Its patent-pending system architecture and custom-designed ASIC provide the most secure solution for critical enterprise data.
Firewall attributes include:

- Network address translation (NAT) hiding inside IP addressing
- Transparent mode, which is invisible to network devices
- Dynamic filter, to protect network services
- User authentication to allow authorized access only

Its design supports wireline performance for critical VPN functions and sets the standard for VPN solutions that include:

- IPSec-compatible interoperability with other network devices
- IKE key management, which secures the exchange of keys
- DES and triple DES, which offer the highest possible level of encryption
- Strong authentication, offering MD5 and SHA-1

Traffic shaping enables network administrators to monitor, analyze, and allocate bandwidth to support network traffic. It offers:

- Traffic bandwidth and prioritization management to manage traffic in accordance with user expectations
- Load balancing, to support server farms
- Virtual IP maps and protection for internal servers

NetScreen-100 provides real-time monitoring and logging of network traffic, offering:

- Real-time logging and alarming, ensuring network status monitoring
- SYS Log, which integrates with existing monitoring systems
- E-mail alerts to notify administrators of system and network events

NetScreen-100 can be configured remotely using universal browsers or locally with a Windows 95/NT utility, SNMP manager, or command line interface. In more detail:

- Web support offers a popular browser management interface
- Windows 95/NT GUI supports local secure administration
- SNMP management capabilities
- Command line interface allows script and modem control

8.5.7   PACKETSHAPER FROM PACKETEER

Packeteer was formed to address the crowd control problems caused by the runaway growth of the Internet and its adjunct, the World Wide Web. This growth has created dissatisfaction among end users frustrated by poor quality of service (i.e., slow response time and unpredictable Internet access), corporations unable to exercise precise control over their own networked resources, and Internet service providers unable to deliver differentiated levels of service to their customers. PacketShaper optimizes and controls Internet traffic over wide-area links, letting network managers set — and enforce — policies with regard to bandwidth allocation and prioritization.

Given a population of a few high-speed users and many low-speed users contending for a server with a T1 access link, a reasonable goal would be to give every user a certain minimum committed rate at which information would be transferred and allow for an even division of excess bandwidth on demand. Also critical is providing the network manager with visibility into how often users find their Web servers stuck — due to congested links, poor server performance, ISP infrastructure problems, etc. — and abort their transactions in disgust.

8.5.7.1   TCP Rate-Based Flow Control

The aim of the product is to enable network managers and ISPs to manage their piece of the Internet — the specific access links under their control. Fundamentally different from queuing-based QoS approaches, Packeteer explicitly controls the rate of individual TCP connections by managing end-to-end TCP flow control from the middle of the connection; this direct feedback to end systems smoothes out the normal business of the TCP traffic, avoiding retransmissions and packet loss and delivering a quality of service perceived as consistent by the user. The benefits of this approach are realized by both enterprise network managers and ISPs. Network managers can establish policies for explicit allocation of committed and excess bandwidth, maximizing the number of users accommodated on a given access link and improving each user's quality of service. Packeteer has devised a method of detecting the access speed and network latency of remote users in real time, allowing resource allocation decisions to be made with information about the potential data rate of individual connections. In contrast to routers, which cannot manage incoming bandwidth, the Packeteer solution controls both inbound traffic (e.g., Web site access) and outbound traffic (e.g., Web browsing from corporate intranet users).

With the ability to control the mix of different traffic types that characterizes every enterprise network, network managers can cope with the new generation of "push" applications (e.g., PointCast, Marimba Castanet) that have contributed significantly to network congestion and slowdown. They can "push back" at these applications, assigning them a low priority; thus mission-critical data such as e-mail, electronic commerce, or Web site inquiries are guaranteed to get through, while PointCast, for example, is transmitted at a slower rate using excess bandwidth.

ISPs hosting Web sites for many customers can now set access policies, ensuring that "bandwidth hogs" — bandwidth-intensive sites that typically use extensive graphics or animation and frequently download large files — share the available bandwidth. Furthermore, Packeteer technology lets ISPs establish different tiers of service, based on bandwidth guarantees, for which they can charge different rates. They can also match users' requests for content to the Web-based service most appropriate for their connection speed; for example, low-speed users may be directed to text or concise graphics, whereas high-speed users can access rich, data-intensive images.

8.5.7.2　PacketShaper IP Manager

The PacketShaper family of products are IP bandwidth management devices, typically placed on the network between an access router or access concentrator and a LAN. They transparently manage bandwidth, report on the user's experience, and match content delivery to connection speed. PacketShaper is a passive network element that operates independent of the system and requires no changes to, or reconfiguration of, the TCP/IP network or end stations.

Figure 8.7 shows PacketShaper in operation. It resides behind WAN devices, enabling network managers to control bandwidth for inbound and outbound traffic flows.

Two scenarios in which the benefits of PacketShaper are readily apparent are corporate WANs and ISP Web hosting services. In a typical geographically dispersed corporate environment, access to the headquarters-based central data bank from a remote regional office can be guaranteed a minimum required rate; casual Internet

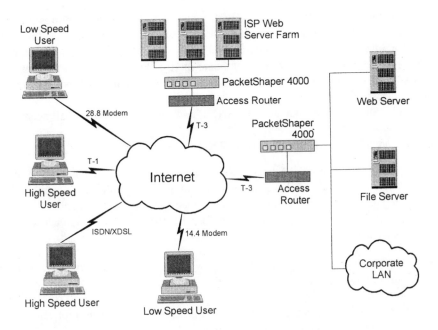

FIGURE 8.7   PacketShaper from Packeteer in operation.

access and Web surfing is allotted only the excess bandwidth. In an ISP Web hosting scenario, different bandwidth levels can be explicitly allocated to co-located servers; one excessively used site will not dominate available bandwidth and negatively impact service at the other sites. PacketShaper provides bidirectional TCP rate control, permitting users to set priorities by user and/or applications and to allocate bandwidth according to those priorities. Bandwidth allocation can be fixed or dynamic based on the network load. PacketShaper also reports on the metrics of the user experience. Problems are isolated by server, access link, and service provider, allowing the network manager to pinpoint a problem and speedily resolve it. Reports are delivered via the browser to any station on the network.

PacketShaper's technology differs from RSVP (resource reservation protocol). RSVP was designed to reserve bandwidth for inelastic, real-time network traffic. Inelastic traffic does not tolerate delay conditions — that is, certain applications require a minimum amount of bandwidth within a specific timeframe. RSVP is intended to solve problems with multicast sessions — that is, audio and video streams on backbones. This protocol requires end-to-end RSVP deployment, which means that end systems and routers require updates so they can communicate reservation requests. Even if the whole network is RSVP-aware, there is no policy server or arbiter to apply network-wide rules for priorities and resource allocation. RSVP effectively supports applications such as video conferencing, but in its current form, it won't have any impact on Web experience. It is designed for a small number of persistent flows, whereas Web traffic is characterized by a large number of transient flows.

PacketShaper enables control of both inelastic traffic, which requires a constant rate, and Web traffic, which tends to be bursty. Packeteer addresses Internet QoS for interactive services, such as Web browsing. Most Web traffic does not require a constant bit rate. To accommodate the totally different data types, such as text versus 24-bit graphics, it is very important to smooth the bursty nature of the Web.

PacketShaper can be managed by SNMP and can incorporate SNMP agents.

8.5.8 WEB SERVER DIRECTOR FROM RND

With the incredible growth of the WWW, popular Web sites are being overwhelmed by traffic. Web site servers have limits to the number of simultaneous active users they can handle. To overcome this limitation, Web site managers can either increase the capacity of the server or create a server farm by using multiple redundant servers in one site. Both solutions have their drawbacks:

1. Replacing the server with a more powerful server is very expensive and offers no real redundancy because:
   - Stronger servers are much more expensive.
   - This is not a scalable solution and there is no investment protection.
   - There is a single point of failure.
   - Service is interrupted during maintenance and upgrade periods.
2. A server farm is a more cost-effective, fault-tolerant solution but also is inefficient for the following reasons:
   - Each server must have its own IP address, which is seen by Internet users.
   - Users have a more complicated interface with the Web site. They have to choose a specific server; thus, the load distribution is random and not efficient.
   - Users can overload a certain server, getting no response, while another server is idle.

Web Server Director (WSD) from RND helps network managers build multiple WWW sites that appear to the user as a single address. Only the IP address of the WSD is necessary for access to the WWW. The WSD distributes the traffic between the servers using a sophisticated load balancing algorithm. The Internet service provider can connect or disconnect servers to the site in a manner completely transparent to the users. Servers can be connected locally and even distributed effectively with unlimited capacity.

The Web Server Director is available with two or four Ethernet or fast Ethernet ports. One port is connected to the Internet access router; the remaining ports can be connected to the servers. Each port can handle one or more servers, and several WSDs can be cascaded in the local site and even between remote servers using remote access nodes: VAN or R-dapters as WAN gateways for leased lines, PSTN, and ISDN services.

Web Server Director provides the following features:

- Sophisticated load distribution among multiple local and remote servers using one of three available user-selectable distribution methods
- Server fault detection allows for optimal server utilization. The server can be connected and disconnected from the site in a manner completely transparent to the users. An inactive server is automatically extracted from the distribution list, the WSD directing all the traffic to the current active servers. As a server becomes active again, it is automatically inserted into the distribution list.
- Up to 100 servers can be configured. Priority can be given to each server, allowing for more efficient and optimal load distribution while directing more traffic to stronger servers.
- Two WSDs can back up one another, allowing for a completely fault-tolerant system with no single point of failure — multiple redundant servers and redundant Web Server Directors.
- Smooth and transparent installation, with no need for network reconfiguration
- Scalable solution, which provides investment protection for growing Web sites
- The clients table is dynamically learned, containing the current active users and their connection time, allowing for dedicated and useful statistics as well as monitoring.
- Access control up to the application layer provides an additional layer of security.
- SNMP management can be achieved via one MultiVu platform: HP 9000, RS 6000, HP OpenView for Unix and HP OpenView for Windows, and ConfigMaster, a Windows-based management platform.
- Various statistics can be obtained such as current server load (in packets), current attached clients per server, number of unsuccessful attempts to the WSD due to clients table overflow, etc. Traps will be initiated in the case of special events such as clients table overflow, server status change, etc.

Principal applications include:

1. Directing WWW traffic — The Web Server Director is ideal for a local Web site or for redundant servers as well as for distributed server farms. Connecting cost-effective remote access nodes, VANs or R-adapters from RND for slow links and the MRT for 2-Mbps links allows for a distributed solution using all available WAN services and features. Figure 8.8 shows a local solution and Figure 8.9 a distributed solution.
2. Flexible addressing — Server farms can include different servers with various processing power. With WSD, stronger servers might require a full 10 Mbps, while others can share the same Ethernet. Either way, the

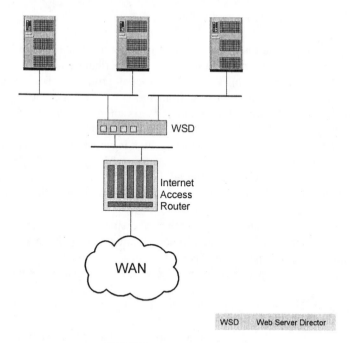

FIGURE 8.8   Local solution.

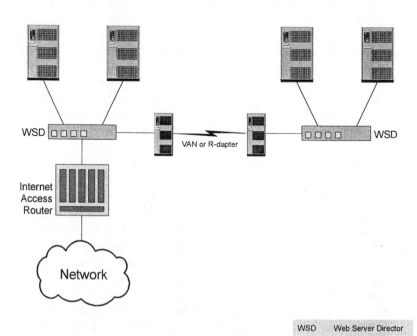

FIGURE 8.9   Distributed solution.

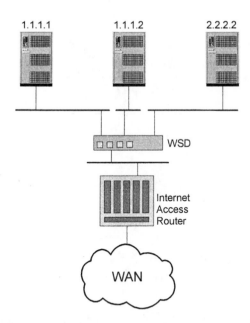

FIGURE 8.10   Solution with dedicated ports per server.

addressing of the servers is flexible. Servers can be given different or identical IP subnets and can all reside on the same IP subnet regardless of their physical location. Figure 8.10 shows a solution with dedicated ports per server. Figure 8.11 displays a solution with multiple servers on the same LAN.

3. Avoiding a single point of failure — The Web Server Director provides a full tolerant network, ensuring maximal functionality of the Web site, as shown in Figure 8.12.

8.5.9   IPATH-10 AND IPATH-100 FROM STRUCTURED INTERNETWORKS, INC.

Products from Structured Internetworks give network system administrators, Webmasters, and Internet service providers the tools to provision and guarantee levels of service, prioritize server resource access, and deliver consistent performance under any traffic load conditions.

Active traffic management for IP networks means the end of sporadic, uncontrolled data flows and runaway traffic from one user or application blocking access for others. IPath active traffic managers are a family of products that allow Internet access points, commercial based or private net based, the ability to provision, guarantee, and dynamically adjust the flow of data through primary network connections to critical server resources or users. The products are typically positioned after a gateway router and before a local network to dynamically intercept and adjust incoming and outgoing traffic flows, to bind them to administrator-defined policies for minimum guaranteed and maximum burst transfer rates on a per host, network,

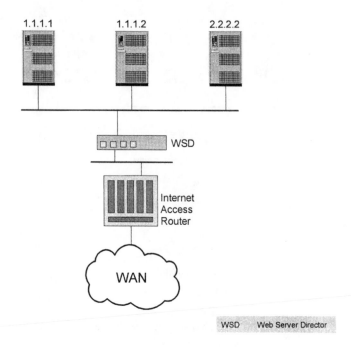

FIGURE 8.11 Multiple servers on the same LAN.

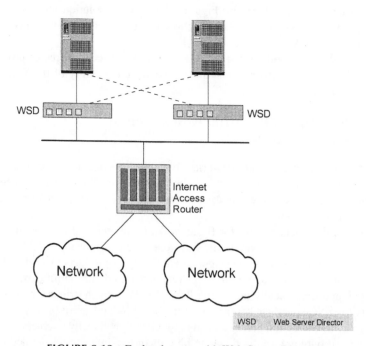

FIGURE 8.12 Fault tolerance with Web Server Director.

or application-type basis. The device does not compete with or replace other core networking devices, such as routers, switches, or servers. Instead, it complements these devices, supplementing their connectivity, transport, and content provisioning functionality with advanced traffic prioritization, allocation, and administrative management services.

Key features of the IPath products include:

- Active throttle-back rate shaping technology, which controls bandwidth congestion at the source
- Bandwidth allocation with guaranteed minimum bandwidth partitioning
- Dynamic load distribution for maximum link utilization
- Oversubscription of total link bandwidth, with individually specified burst rates
- Individual host IP address and application support, along with subnet and host grouping
- Bandwidth allocation resolution in 2-Kbps increments
- Highly reliable, no moving parts design, including fault-resilient bypass circuitry for uninterrupted data flows
- SNMP MIB-II support for statistical information access and high-level manager interfacing
- Statistical and graphical traffic accounting for capacity planning and charge back

The scalable design supports cost-effective, plug-and-play integration anywhere networks require increased predictability of traffic flows and managed access to critical server resources — from heavily congested shared access links to the Internet or corporate WAN resources, to dense intranet or Web server sites.

8.5.9.1 IPath-10 Active Traffic Manager

This is a high-performance custom processor-based Ethernet device that provides real-time network bandwidth management, allocation, and accounting for up to four 10-Mbps segments. Positioned in front of key network segments, it dynamically structures incoming and outgoing IP connections into policy-defined behaviors of

- Minimum bandwidth reservation
- Maximum burst throughput rate
- Groupable bandwidth classes

Policies are definable for hosts, subnets, or groups of hosts and subnets. For finer granularity, bandwidth policies are also definable for IP services and applications.

This product can control and monitor network bandwidth across as many as 2000 defined hosts, groups, subnets and Internet services. While the IPath Manager™ and IPath Command Shell™ simplify the process of dynamically configuring and finetuning resources, the software suite also displays and exports historical traffic usage in real time or near real time in HTML format by bandwidth policy, for

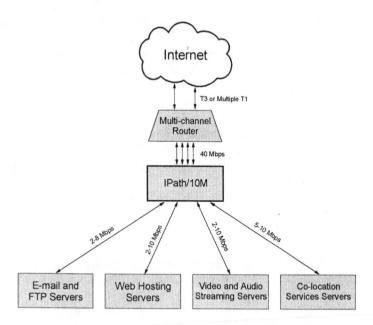

FIGURE 8.13   Location of IPath-10 active traffic manager.

capacity planning and accounting. These tools enable easy control and maximum use of available network resources. Figure 8.13 shows the location of the IPath-10 active traffic manager.

8.5.9.2   IPath-100 Active Traffic Manager

In addition to other features of the low-end product, wire-speed 100-Mbps performance provides the solution for guaranteed and segmented bandwidth to Internet and shared network services.

This product can control and monitor network bandwidth across as many as 2000 defined hosts, groups, subnets, and Internet services. IPath Manager™ and IPath Command Shell™ simplify the process of dynamically configuring and fine-tuning resources to hosts, while IPath Profiler™ logs and presents trending information in standard database and HTML formats. The IPath Manager enables the easy management, coordination, and validation of policies through a common policy server to support multiple IPath devices and/or departmental administrators on larger, enterprise networks. The software suite also displays and exports historical traffic patterns and usage data, for capacity planning and accounting, to allow service providers and network managers the ability to easily control and maximize available resources. Figure 8.14 shows the location of the IPath-100 active traffic manager.

8.5.9.3   IPath Profiler

The IPath Profiler monitors and reports traffic usage for any IPath-defined bandwidth policy by generating HTML Web pages containing utilization graphs. These pages

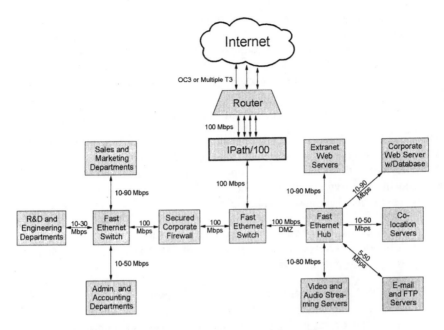

FIGURE 8.14   Location of IPath-100 active traffic manager.

provide a near real-time visual representation of traversed and historical traffic patterns. IPath Profiler for Windows NT uses a full graphical interface for configuration. It can also graph the traffic observed over days, the last week, the last 4 weeks, and the last 12 months. The IPath Profiler records into a database all the collected data from IPath and consolidates it over time, maintaining all relevant data for up to 2 years.

The most important features of the product suite include:

- Definable minimum and maximum bandwidth reservation thresholds
- Dynamic allocation of available bandwidth to active users via burst mode oversubscription
- Application-specific hardware for line rate processing with submillisecond delays
- Direct TCP throttling to rate control traffic with no packet loss or retransmissions
- Internet service and application-addressable bandwidth control
- Subnet and host-definable grouping and oversubscribing
- Multichannel coordinated or independent bandwidth control
- SNMP manageable through MIB-II and private MIBs
- High reliability through fault-tolerant bypass circuitry and comprehensive "power on self testing" (POST)
- Nonregistered IP and protocol support
- IPath Manager for NT and Java, enables the flexible definition and deployment of bandwidth policies through simple, user-definable templates

- IPath Profiler gives instant access to statistical and trending data from any universal Web browser
- IPath Command Shell allows quick terminal or telnet access via an intuitive command hierarchy
- Remote management via modem

8.5.10 BANDWIDTH ALLOCATOR FROM SUN MICROSYSTEMS

Sun's Bandwidth Allocator enables Internet service providers and enterprise IT departments to perform bandwidth provisioning and accounting to help ensure quality of services to their customers. IT departments need to be able to guarantee their users quality of service, and ISPs need to offer customizable service level agreements. To do this, they must be able to:

- Provide guaranteed bandwidth and quality of service
- Monitor the levels of bandwidth and quality of service they are providing
- Keep corresponding accounts

Bandwidth Allocator is a software product that provides the means to perform all these actions. By installing it on major links and known congestion points of networks — and by setting consistent policies — network managers can attain bandwidth control throughout the network.

The structure of the product is shown in Figure 8.15. It is based on a STREAM module, which resides between the IP layer and the network interface. The packet classifier takes any outgoing packet and assigns it to a traffic class according to information in the IP or application header. The rules governing the assignment of quality of service are taken from a provisioning rules file. The provisioning rules file can be updated by command-line mode or via a GUI. The different classes of traffic are transmitted by the packet scheduler according to the predefined priority and transfer rate for each class. Information collected during the classification process is available to accounting through the GUI or via a command line interface. The Bandwidth Allocator can be monitored via SNMP; the current configuration and statistics are available through a proprietary MIB.

Key benefits of the product include:

- By enabling control of the bandwidth allocated to users, applications, and organizations that share the same link, Sun Bandwidth Allocator provides the means to enable service providers to deliver adequate levels of service without overprovisioning their network equipment.
- The traffic prioritization by Sun Bandwidth Allocator considerably reduces the risk of network congestion and prevents a small number of applications or users from consuming all the available bandwidth.

Bandwidth Allocator controls traffic sent over a link. It can be installed in two environments: as a traffic manager or as an application performance manager. As a traffic manager, it can be installed in IP-transparent mode on a device that controls access to the network (LAN, WAN, or Internet). Used in this way, it controls traffic

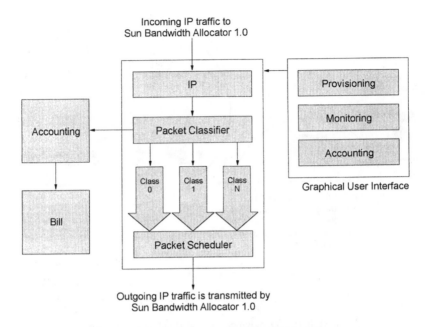

FIGURE 8.15   Structure of Sun Bandwidth Allocator.

while remaining transparent to IP users. The IP traffic is sorted and prioritized by application, traffic type, or customer. Installed as an application performance manager, it controls IP traffic from a server to the network (LAN, WAN, or Internet). The server may be a file server, a Web server, or any application server. Traffic can be controlled by the application and/or by the customer. Figure 8.16 shows how Bandwidth Allocator is deployed.

There are three principal functions of the Bandwidth Allocator:

1. Provisioning by rule enforcement — It manages traffic transmission based on provisioning rules that sort and prioritize traffic according to:
   - Traffic type (e.g., HTTP, FTP, e-mail, news, telnet, or NFS traffic)
   - End user source or destination address
   - Network source or destination address

   Some examples of specifications that can be defined using provisioning are as follows:
   - An ISP with a single connection point to the Internet can restrict FTP traffic for Customer A to no more than 5% of its bandwidth and limit Customer B to no more than 20% for Internet access.
   - An ISP can reserve a minimum of 60% of traffic outgoing from the Web server for HTTP, restrict FTP traffic to 10%, and leave up to 30% for its news service.
   - An IT department can reserve 50% of a WAN link to traffic that originates from machines located in offices A, B, and C. Then, within this 50%, it can reserve 10% for NFS traffic.

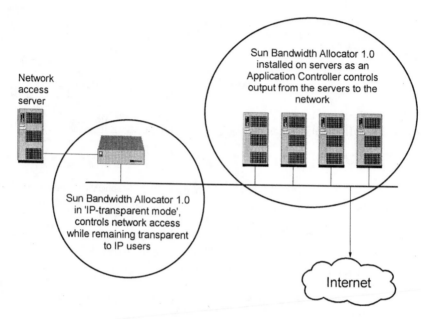

FIGURE 8.16   Deployment of Sun Bandwidth Allocator.

- The product can ensure that FTP transfer between machine A in office X and machine B in office Y will not use more than 30% of the capacity of the line between the two offices. If there is no FTP traffic, other types of traffic could borrow the available bandwidth.
2. Remote monitoring — The product provides real-time statistics on resource usage. These can be accessed via a Java-based GUI, Solstice Enterprise Manager, or any SNMP manager. A statistical API enables customers to integrate Sun Bandwidth Allocator statistics into their own monitoring systems.
3. Web flow accounting — Flexible accounting schemas allow payment by class of service per customer or by actual bytes or packets transferred. A Web flow agent collects information and outputs it in ASCII format, which can be automatically sent to a billing system.

Key features of the Sun Bandwidth Allocator are as follows:

- Bandwidth management rules can be configured to map the organization, systems, or geographical layout.
- The product manages any type of IP-based traffic, is transparent, and works within a heterogeneous environment without any modification of the systems accessing the gateway.
- A comprehensive and user-friendly Java-based configuration utility makes it easy to specify bandwidth allocation policies and perform remote management from anywhere on the network.

- Reporting utilities can be used to monitor network use by traffic type and IP address.
- The product runs over WAN and LAN links such as Ethernet and FDDI. It can also be integrated with Web servers to provide outgoing flow control.

The product requires a SPARC or Intel platform with a minimum of 32 MB of RAM and 21 MB of disk space as well as the Solaris operating system.

8.5.11 TrafficWARE from Ukiah Software

NetRoad TrafficWARE provides policy-based, directory-enabled IP traffic management, including both monitoring and QoS control. It is designed to solve performance problems in IP-based networks, especially those dependent on wide-area communication. It delivers better network quality of service, improving network response time as well as cutting both WAN costs and network cost of ownership. Because demand for bandwidth over WANs will outstrip supply for the foreseeable future, the managed approach to allocating bandwidth and prioritizing traffic makes a lot of sense. In addition, the product makes traffic management easy, by using a policy management approach that is directory enabled so that traffic rules can be based on network users and groups. There are three products in the TrafficWARE family:

8.5.11.1 TrafficWARE Gateway

This product manages IP network traffic at network bottlenecks such as the WAN (Internet or intranet) access point or a server farm access point. It provides a powerful system for policy-based traffic management and can be integrated with directory services such as NT domains and NDS to enable traffic rules based on users and groups.

Key features of the product include the following:

- Monitors and controls IP network traffic to improve performance for critical users and applications
- Provides real-time and historical views of network traffic and integrates with NT's performance monitor
- Enables traffic control through bandwidth allocation, traffic prioritization, and admission control
- Supports IP types of services to enable standards-based traffic control
- Application-level intelligence provides support for managing network traffic for business applications
- Provides comprehensive alarms
- Support for throughput up to 100 Mbps, with extremely low latency
- Easy installation with transparent gateway — no subnetting required
- Installs on NT servers, which may also be running NetRoad FireWALL from Ukiah

8.5.11.2   TrafficWARE AS

This product solves network performance problems at the source — on the application server. It can be installed on an NT 4.0 application server running other applications, such as Microsoft IIS, MS Exchange, Citrix MetaFrame, or Lotus Notes/Domino. It can be installed on all the critical NT application servers throughout the network. When installed on an NT application server, TrafficWARE AS can monitor and control the IP traffic to and from that server, regardless of whether the traffic is LAN or WAN oriented. TrafficWARE AS's ability to integrate with the NT performance monitor is particularly useful on an application server, providing an integrated view of both system and network performance information.

Key features of the product include the following:

- Monitors and controls IP network traffic to improve performance for critical users and applications
- Provides real-time and historical views of network traffic and integrates with NT's performance monitor
- Enables traffic control through bandwidth allocation, traffic prioritization, and admission control
- Supports IP types of services to enable standards-based traffic control
- Installs on NT 4.0 servers, which typically are running other NT applications such as Web servers, databases, and mail servers
- Provides comprehensive alarms

8.5.11.3   TrafficWARE Monitor

This product is extremely useful for network troubleshooting. It provides a user-centric and application-centric network monitoring tool, using technology common to the rest of the TrafficWARE line. It can be deployed anywhere on the network, such as at a WAN gateway point or on an NT application server. It is a valuable complement to the active traffic shaping features provided by TrafficWARE AS and TrafficWARE Gateway. It is also the first in the family to attempt policy-based traffic management.

Key features of the product are as follows:

- Monitors and controls IP network traffic to improve performance for critical users and applications
- Provides real-time and historical views of network traffic and integrates with NT's performance monitor
- Provides comprehensive alarms
- Installs on NT 4.0 (workstation or server), which may be dedicated to this application or may be running other applications

The three products can be used on their own or together in the same network. When all three products are used, TrafficWARE provides the most complete, policy-based, distributed traffic management system. Deployment choices include:

- At the network edge behind the WAN router (TrafficWARE Gateway or Monitor)
- In front of a server farm (TrafficWARE Gateway or Monitor)
- On an NT server running an application such as a Web server (TrafficWARE AS or Monitor)

8.5.12 Access Point from Xedia

This product is the first in a new class of Internet access products designed to deliver the traffic shaping and management controls needed to meet service level commitments for a more demanding community of business Internet users. Whether it operates as customer premises equipment (CPE) or as the access bandwidth control point for a managed bandwidth service, Access Point leads in the service flexibility, performance, and ease of use required to build this new generation of business-centered Internet services.

Network service providers can now migrate their customers from T1 to beyond T1 to broadband access rates with a solution that delivers service level commitments and unique management controls to both network service providers and their customers. With Access Point as the customer premises extension of a business class Internet access service, network service providers can now offer broadband Internet access with the operational ease and simplicity they have come to appreciate in their T1 service offerings. The added ability to specifically allocate access bandwidth across departments, applications, and users allows users to creatively define new broadband access services for the enterprise, for multitenant facilities, or for content providers that need to manage shared access to key business services.

8.5.12.1 Managed Bandwidth Services

Network service providers can further leverage Access Point to deliver committed or rate-controlled bandwidth to their Web hosting, server co-location, or downstream remote access subscribers. With Access Point as part of a server farm or a regional point of presence (POP) infrastructure, network service providers now have the ability to offer bandwidth class of service as part of their business services profile.

8.5.12.2 High-Performance Traffic Control

Although router-only architectures have satisfied basic Internet connectivity requirements, the access landscape is changing. As business customers migrate new applications onto the Internet they need service level commitments to ensure their business objectives are achieved. Routers have attempted to meet this need with early stage priority-based and weighted fair queuing solutions, but these solutions are often delivered at the expense of forwarding performance. Class-based queuing (CBQ) makes Access Point a new class of Internet access solution unique in its ability to deliver advanced traffic management control while also providing high-performance routing for direct WAN connectivity. This single box integration provides a cost-effective migration to high-performance bandwidth-controlled Internet access.

Access Point offers a fully featured IP routing capability, supporting static routes, RIP, Open Shortest Path First (OSDPF), and Border Gateway Protocol (BGP). In addition, the modular Access Point software architecture allows future rapid integration of additional routing protocols and services, such as RSVP, based on customer needs.

Key Access Point product features include:

- Open, vendor-independent IP class of service
- High-speed bandwidth-controlled access to the Internet
- Class-based queuing (CBQ) traffic shaping and control
- Customer and network service provider monitoring and control options
- Ease of installation, setup, and operation

8.5.12.3   Class-Based Queuing

Access Point's CBQ combines high-performance traffic shaping and management controls with a complexity-flexible approach to allocating bandwidth across a hierarchy of traffic classes, where each class represents an aggregation of traffic or an individual connection. Traffic classes are defined based on IP address, protocol, applications, or even URL, providing an intuitive and natural extension of the way administrators currently manage their IP network environments. The unique ability of CBQ to classify traffic as part of a hierarchy allows an operator to first partition bandwidth across multiple departments within an enterprise, and then further allocate bandwidth to individual applications or users within the department. A network service provider can also use this hierarchy to commit bandwidth to each server within a server co-location facility, and then further control bandwidth rates to individual applications or users requiring access to an individual server. Figure 8.17 shows a flexible hierarchy for classifying traffic.

8.5.12.4   Bandwidth Borrowing

One of the most powerful features of CBQ is bandwidth borrowing. Borrowing allows a traffic class to request access to more than its committed bandwidth to meet periodic burst in demand. Borrowing privileges can be established for any individual traffic class, providing a completely flexible mechanism for service subscription. With bandwidth borrowing, a network service provider can actually define different class of service levels that create new revenue opportunities. Borrowing can also become an important planning tool, because Access Point statistics, maintained for each traffic class, allow a network service provider to monitor borrowing activity and anticipate the need for bandwidth beyond the currently committed rate. CBQ bandwidth borrowing is shown in Figure 8.18. A typical dialog is also included.

8.5.12.5   Management and Administration

Access View provides a comprehensive suite of tools for configuration and monitoring of Access Point. This management solution allows an administrator to easily control Access Point from a graphical Web management navigator, an intuitive CLI,

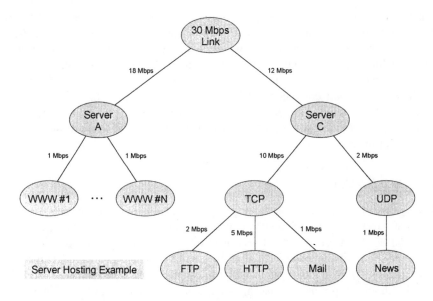

FIGURE 8.17   A flexible hierarchy for classifying traffic.

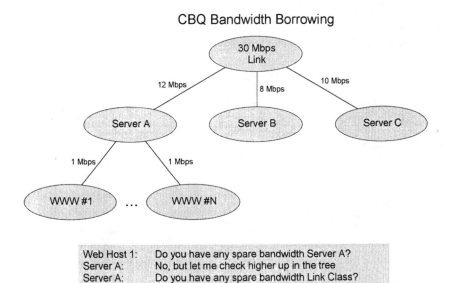

FIGURE 8.18   Class-based queuing bandwidth borrowing.

or any industry-standard SNMP manager. Whereas the Access Point CLI establishes a new standard for ease of configuration and management, the Web interface provides a powerful graphical tool for continuous monitoring of bandwidth allocation and usage. The Web management navigator provides an embedded Web server, allowing any Java-enabled Web browser to navigate through a simple, forms-based configuration manager. Java applets served directly from Access Point also provide a real-time graphical display of key statistical information, including current bandwidth allocation and usage. CLI ensures ease of use with the extensive use of defaults and intelligent command completion. The CLI is built on top of the tool command language (TCL), which provides a powerful scripting tool for automating routine or bulky management tasks. Access to the CLI is via an in-band telnet or out-of-band VT-100 terminal session.

8.6 SUMMARY

Load balancing is continuously gaining importance. Solutions are partially based on hardware, partially on software. Both alternatives have benefits and disadvantages. This industry is just emerging. Users need time to collect experiences with functionality, scalability, performance, security, ease of installation, documentation, and manageability of load balancing products. Most likely, these products will be combined with server optimization packages to give users a more complete and integrated solution.

9 Look-Through Measurements

CONTENTS

9.1 INTRODUCTION

Web application requirements have gone from zero to mission critical within a very short period of time. The available tools have not kept up with this pace. In a business environment where "connections failed" means the same thing as "closed for business", IS/IT professionals are left to struggle with the challenges of building a highly available, high-performance intranet infrastructure.

There are many problems that interact with each other:

- The majority of Web sites, both Internet and intranet, use single Unix or NT servers. Like mainframe solutions of the past, these centralized servers have become single points of failure. Even minor system upgrades become major service problems for demanding users.
- As the demands of interactivity grow, the cost of WAN bandwidth becomes a major factor. System configurations that force all user access out across the WAN for each request stretch out retrieval times and raise users' frustration levels.

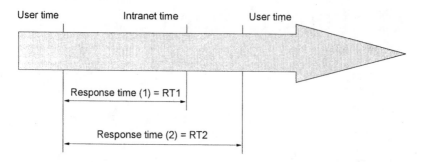

FIGURE 9.1   Response time definitions.

- The increasing complexity of Web applications adds even more overhead; electronic commerce and multi-tier content architectures that build pages on the fly out of applications and databases make high reliability an even more important — and more costly — goal.

The most severe problem is that the Web technology base is narrow. In other words, the solutions that can be applied to these problems are expensive and not very effective. Adding WAN bandwidth and a bigger server is just the first step in a never ending circle. Adding mirrored, distributed servers increases server costs significantly as well as the complexities and costs of content distribution. Hiring more Webmasters and site administrators to reboot downed Web applications and servers is not the ultimate solution. And in a world of increasingly dynamic content and transactions, how effective will server caches and load balancing tools be?

Various tools have been referenced in the previous chapters. This chapter takes another approach, using the end user perspective to measure, interpret, and report service quality metrics.

9.2   RESPONSE TIME MEASUREMENTS

Response time is one of the key metrics in all service level agreements. Its definition varies, but users usually consider it to be the elapsed time between sending the inquiry and receiving the full answer. Figure 9.1 displays the differences between two alternative definitions:

- Time up to the first character of the response on the screen of the user
- Time up to the last character of the response on the screen of the user

The second definition is better suited for the working cycle of users. The difference between RT2 and RT1 depends on many factors, such as the throughput of the backbone and access networks, the servers in these networks, the number of hops, and the hardware/software capabilities of the clients' workstation or browser. Present measurement technology offers the following alternatives (Jander, 1998):

- Monitors and packet analyzers: They filter and interpret packets and draw inferences about application response times based on these results. These monitors passively listen to the network traffic and calculate the time it takes specific packets to get from source to destination. They can read the content of packages, revealing eventual application errors and inefficiency. But, they cannot measure response time end to end.
- Synthetic workload tools: They issue live traffic to get a consistent measurement of response time on a particular connection in the intranet or for a given application. These tools are installed on servers, desktops, or both. They typically send TCP messages or SQL queries to servers and measure the time elapsed until the reply. Results from multiple sources are correlated to give a more detailed view about intranet response times. They are very accurate about end-to-end response time.
- Application agents: They work within or alongside applications, using software that monitors keystrokes and commands to track how long a specific transaction takes. They can run at both the client and the server. They clock specific portions of the application at the server or the workstation. The use of agents needs customization and the correlation of many measurements to give users a performance estimate about their intranet.
- ARM MIBs: ARM defines APIs that allow programmers to write agents into an application so that network managers and Webmasters can monitor them for a range of performance metrics, including response time. It is a complete offer to application management by existing management platforms. However, it requires that the customer rewrite existing code, which many companies are unwilling to do.

Figure 9.2 shows the locations of these tools and agents.

Administrators must consider the following factors when evaluating products:

- Customization needs
- Maintenance requirements
- Deployment of code
- Overhead of transmitting measurement data
- Load increase due to synthetic workload
- Reporting capabilities
- Capabilities to solve complex performance problems
- Capabilities to conduct root-cause analysis
- Combination with modeling tools
- Price of the tools

9.3   HIGHLIGHTING BOTTLENECKS

End-to-end service level monitoring is gaining popularity with Web-based applications. Monitoring refers to targeting availability and response time measurements. Element-centric management platforms "look down" and manage elements. Response time monitoring tools "look through" the infrastructure from one end to the other.

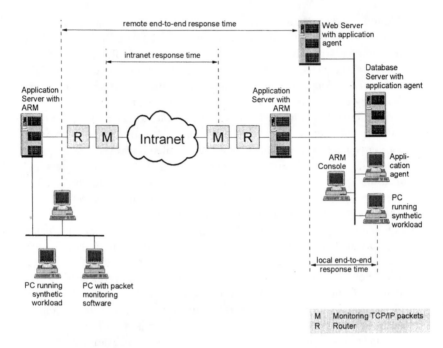

FIGURE 9.2   Positioning response time measuring tools.

Application-related measurements can also be done with RMON probes. The way to do this, according to NetScout Systems Inc., is to track an application on its entire path across the enterprise. To support that approach, the remote monitoring vendor is able to collect and report traffic statistics IT managers use to measure how quickly an application makes its roundtrip run.

NetScout is leading off its "application flow management" strategy with a new multiport fast Ethernet probe, an RMON2 agent for NT servers, and Web-based reporting software. Applications can be observed and measured as they run using AppScout, a browser-based solution. AppScout monitors SAP R/3, Microsoft Exchange, Lotus Notes, and TCP/IP applications.

Typical look-through products work on the principle of Java applets in combination with C++ scripts. The code is distributed to various selected end points in the network. These agents generate synthetic transactions against targeted applications, such as databases or intranet Web pages. Response time for these scripted transactions — including the response times over each individual "hop" along the route — are logged on a management server, which assembles and organizes the collected data. The data is then available to users through a client-side Java interface.

The new type of network instrumentation closely mimics the end users' actual experience because it measures the end-to-end responsiveness of an application from one or more outlying LAN nodes to the application server and back again. By doing so, it delivers a metric that accurately reflects application and service performance levels on the network. Attempts to gauge the end-to-end performance level of an application over the network by monitoring each distinct element along the service delivery path have proven unsuccessful. Element-specific monitoring is still essential for troubleshooting and maintenance, but network managers have to look at new kinds of instrumentation if they want to view the environment from the end user's point of view.

9.4 TOOLS FOR LOOK-THROUGH MEASUREMENTS

A couple of emerging tools measure the end user response time. The techniques these tools use might be different, but the ultimate goal is the same: to get a close assessment of how end users experience intranet performance. Each tool tries to correlate performance analysis results and business applications.

All the products reviewed in this section support real-time measurements, troubleshooting, application impact analysis caused by traffic congestions, and response time sampling by synthetic workload.

9.4.1 SERVICE MANAGEMENT ARCHITECTURE FROM JYRA RESEARCH

The need for service management as a complementary management architecture has been driven, in part, by the difficulties encountered in delivering element-oriented management services. Furthermore, it has also been recognized that, for monitoring network service management across multiple network domains, an elementary approach is unsuitable. This product reports on the performance and latency of intranets by monitoring the flow of the application's traffic along its entire length. This makes identifying bottlenecks straightforward — even to the devices — and enables end users to effectively monitor network and outsourcer performance.

Being scalable, the architecture of the product incorporates the following characteristics, all of which are essential in delivering complete service management:

- The resilient, network-wide data collection system does not affect user traffic.
- A distributed analysis capability allows variable values to be monitored across the whole network. Sophisticated analysis can therefore be carried out at remote locations or even from the desktops.
- Files can be transported back for analysis into the management centers.
- It allows browser-based access to network-wide statistics, from any point in the network.

Service management architecture goes further than the traditional element-oriented metrics by measuring levels of service; it therefore has wide commercial applications, such as:

- Managed bandwidth providers can allocate easily monitored service profiles to each user, desktops, applications, and networks. This may then be incorporated into a service level agreement between customers and suppliers. It adds value to customer services by optimizing and employing

capital equipment more effectively. Next, investment can be tightly
focused against end user deliverables, leading inevitably to an improved
return on investment.

- Internet service providers can offer a varying level of service, maximize
  bandwidth efficiency, and offer premium intranet and extranet services.
  Mass market electronic commerce over the Internet can also be supported
  by providing a secure managed platform within this particular environment.
- Corporate users can monitor each network connection, which can then be
  securely and cost-effectively analyzed and resulting trends identified.

The effect of changes to the infrastructure can be evaluated from the end user
experience. Network statistics gathering is fully automated and includes those points
on the network which have been outsourced — those which are the responsibility
of external agencies.

Most conventional network management architectures use the same basic structure and set of relationships. Network monitoring devices, such as RMON probes,
run software allowing them to send alerts when they recognize problems. Problems
are recognized when one or more user-determined threshold is exceeded. Management entities can also poll the RMON probe devices to check the values of certain
variables. Polling can be automatic or user initiated. Agents in the probes respond
to these polls, providing information to a central management database where a
collective picture of a network can be built. There is a finite limit to how much
management traffic a network will sustain before user response times are impacted.
Polling remote devices to capture trends in variable values is bandwidth intensive,
as is transporting significant amounts of collected data back across the network for
analysis. Conventional network management tools cannot scale to offer businesses
true network service measurements for the following reasons:

- Large volumes of uncorrelated management traffic are generated.
- There is minimal correlation between the many disparate management
  systems and the users' desire for measurable network performance.
- The network support team requires a high degree of knowledge.
- Application response time to users is not directly measured.
- The limited scalability of central polling architectures restricts deployment
  of the limited functionality currently available.

Figure 9.3 shows the agent deployment from Jyra Research.

It is possible to design a network service management solution that overcomes
these limitations; however, it needs to be based on a new, more sophisticated,
service-oriented, distributed, management architecture. Systems that rely on polling
remote devices to monitor variables are not practical for monitoring values that
vary constantly, such as application network usage and application response time
statistics.

The solution by Jyra Research has implemented a distributed management architecture that will accommodate large volumes of management traffic and the scale

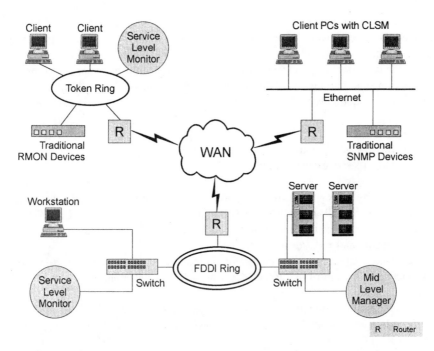

FIGURE 9.3 Jyra software agents deployed independently of platform and network technology.

of modern networks without affecting user response times. It incorporates the following characteristics:

- A resilient, reliable, distributed, data collection system that does not impact user traffic and that directly measures application response times
- A distributed analysis capability that allows for variable values to be monitored network wide and for sophisticated analysis to be carried out at remote locations (i.e., without transporting large files back to the management center for analysis)
- Browser-based access to network-wide statistics from any point in the network

Jyra Research has included many of the accepted network management data collection standards such as SNMP, MIBII, and RMON in the Java distributed language and has provided a means within the architecture to consistently monitor application response times at the edge of the network.

The most important benefits of the distributed response time measurement are:

- Direct and persistent measurement of application response times
- Ability to scale to manage large networks
- Low levels of management traffic
- Broader access to network management information
- Rapid development and deployment of new management applications

Jyra's service management architecture (SMA) has been designed and implemented using the Java programming language. This allows complete platform independence because the code runs initially in a virtual Java machine within its software. The code is written in applet technology, which truncates the main code into compact instruction sets that are easily distributed, yielding ubiquitous flexibility and functionality. It is composed of:

- Midlevel manager
- Service-level monitor
- Client service-level monitor

SMA runs on NT, Windows 95, and Solaris platforms and adheres to industry standards for SNMPv1 and SNMPv2, RMON1 and RMON2, MIB I and MIB II, and Java agents. The requirements for dedicated PC and workstations depend on specific traffic and monitoring environments.

Service level monitors are distributed in the form of lite-Java applets to the edge of a network or even to the users' PCs. The service level monitor

- Measures application response time as perceived by the end users
- Collects SNMP, MIB II, and RMON2 statistics from local devices and RMON probes
- Traces route functionality, which identifies the flow of an application's traffic across a network

A monitor takes advantages of local disk storage to persistently store the collected, time-stamped data locally to reduce the amount of network management traffic generated. Summary collated information is passed upward to allow the network manager to make an intelligent decision regarding viewing the report during the normal working period. Exceptions are automatically generated should a user-defined value be exceeded.

Reports can be accessed by any user from anywhere. Sample reports include the following:

- Line chart report (Figure 9.4): This report shows a trended response time against a marginal and critical level. The line chart can be used to display any application response time required. The gray area behind the line shows the raw data summarized as minimum and maximum values. The line depicts the moving average of the measured response time.
- Trace route monitor report (Figure 9.5): This report details the individual response times for each hop across a large routed network between two sites. Each bar shown in the report represents the minimum, maximum, and average response time for each hop, as well as the sample density of response times during the monitored period. Network managers can quickly and easily show links or routers within the network and determine how these are affecting the users' application response times.

Test Report

/Jyra.com/byadress/IP/176/18/1/176.18.1.254

Hop: IP 176.18.1.254
HopCount: 4

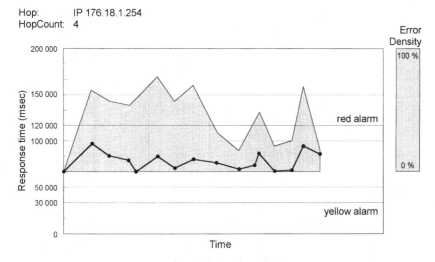

FIGURE 9.4   Line chart report.

Jyra0 - 207.68.137.62 Trace Route Chart

Data gathered: from March 24 / 9:00 AM
                 to March 24 / 10:00 AM

Routes/hops

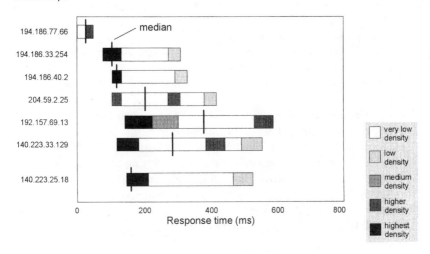

FIGURE 9.5   Trace route monitor report.

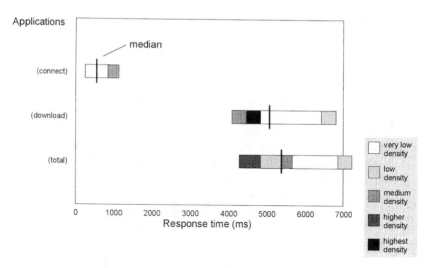

FIGURE 9.6   HTTP application response monitor report.

- HTTP application response monitor report (Figure 9.6): This response chart shows the range of times necessary for a user to connect to a Web site, the range of times required to download a page, and the total time. In this example the Internet (upper bar) performs much faster than the server (middle bar), which accounts for the total time (third bar).

SMA supplies a comprehensive suite of tools and agents that provide the network manager with the following functionality:

- Containerization of network devices to allow business-oriented views of the network
- Network equipment availability monitoring
- End-to-end application response time monitoring and trending of today's client/server-based applications
- Internet/intranet/extranet response time monitoring of HTTP servers and proxy agents
- Detailed end-to-end network latency monitoring and trending
- Network bottleneck/slowlink identification
- Distributed SNMP and RMON1 and 2 data collection with robust and reliable transport of summarized data for reporting
- Distributed relay to provide a customizable and reliable alerting system
- Standard WWW browser-based configuration, control, and reporting (including custom reporting)
- Scalable agent implementation for complete coverage of any sized site or network
- Auto-discovery or import of network topology to ease configuration management

The benefits of this product can be summarized as follows:

- It is specifically focused on the needs of the user. Network managers can monitor the performance of the network with a particular focus on the response time and level of performance experienced by the user.
- It measures the performance levels of WAN service providers with a specific focus on user response times. This allows the network manager to negotiate and measure the deliverables of detailed service level agreements with the service providers.
- It provides scalability. The architecture has been created "from the ground up" as a fully distributed scalable architecture, while focusing on the specific requirements of today's global networks, and yet will scale to accommodate the largest or smallest organization.
- It performs the detailed trending and profiling required to allow full-scale capacity planning, trending, and monitoring, while at the same time allowing the network manager to quickly and easily measure the return on investment of any upgrade or expansion of the network.
- It assists the traditional central poll and collect architecture, creating a next generation distributed data collection system across the enterprise.
- It provides Java software agents of varying intelligence and functionality to allow the network manager to cost effectively distribute the required functionality across multiple platforms and technologies throughout the network.
- It automatically generates detailed but easily understandable HTML reports — for both users of the network and the management chain — showing the overall performance of the network, specifically focused on the user application response times.
- It provides totally browser-based control and reporting that allows access to its reports from anywhere within the network, from any browser on any network station.

9.4.2 NextPoint S3 Performance Monitoring Software from NextPoint Networks

Network device availability is one of the key metrics of service management. This product uses a pointcast-like interface with specific channels to drill down on equipment availability and application response time statistics. NextPoint has extended the product's reach to give IT managers more options for measuring application response times. S3 incorporates a new intelligent agent that records production transaction response times. The same agent also tracks response times of synthetic transactions — a capability bundled with the initial version of the product. That duality gives IT managers the option to customize specific applications so that they can track response times in live production conditions or to simulate IP, Oracle, SAP, and other transactions. To program applications for live-mode response time monitoring, they must be somewhat reworked. That is a task commercial application vendors are still resisting, internal application developers dread, and IT management fears may actually slow down transaction times.

NextPoint offers an API specification, a developers' agent, and agent certification testing. Synthetic transaction monitoring lets IT managers simulate Web, e-mail, domain name service, dynamic host configuration protocol (DHCP), Oracle, SAP, and other application activity to estimate roundtrip response times.

Although the product measures basic response times, it does not have the level of sophistication IT officers may be looking for in the future to aggregate response time information and turn it into a broader measure of productivity.

The product integrates two complementary and important perspectives of the network: the business user and the network administrator. In doing so, S3 answers two important questions: did the network meet business service expectations, and how can service be improved?

Lightweight intelligent agents track business-centric metrics such as response time and availability from strategic locations around the network. Synthetic transactions technology proactively derives end-to-end response times at the application and network levels. The optional agents supplement RMON and SNMP data sources with a more accurate measure of transaction response time from the end user perspective. NextPoint S3 complements the business-centric perspective with comprehensive operational management of the multivendor infrastructure, including real-time alerts, drill-down navigation, and IP services management. It automates real-time fault isolation with policy-based analysis to reduce false alarms. IP services (e.g., DNS, DHCP) represent an increasingly important part of the end-to-end network infrastructure. S3 discovers IP services and monitors their availability and performance. It also automatically identifies duplicate and stale IP addresses. Figure 9.7 shows the service-level management view. The network response time overview is shown in Figure 9.8.

Networks evolve greatly over time, with new users, new applications, and topology changes. Even in their daily traffic variance, however, most networks display remarkable consistency, reflecting the cyclic nature of most business applications. These dynamic patterns may be the result of normal end user and application events such as routine back-ups and employee work shifts. Standard trend analysis helps to understand network traffic variance and to uncover hidden patterns of response time and utilization. Figure 9.9 shows the daily protocol distribution and analysis results.

Web technology is used to distribute information to users. Push technology lets users quickly identify and select information organized by channels of interest, such as service management, exceptions, and applications. Channels are further segmented along resource type, application, reports, maps, etc., so no extraneous information is presented to cloud the message. Users can customize the channels. Information can be accessed from anywhere that is convenient to users. Security mechanisms prevent unauthorized access.

NextPoint S3 provides reports to highlight application and network performance issues, stability problems, and other events on both a real-time and historical

Service Level Management Report

Available Categories:
o   Networks
o   Devices ◄———————— click here to see SLM report
o   Interfaces   for device availability
o   Applications

Application	Source	Destination	Month		Contract		Deviation
			Mar 99				
			IN%	OUT%	IN%	msec	± IN%
WEB	Scan	www.abc.com	0	100	98	1000	-98
TCPPort, Telnet	qatest6	compaq:23	0	100	98	2000	-98
TCPPort, UUCP	qatest6	sunultra:540	100	0	98	2000	+2
Database	tonyp	NpDb/NxPt	40	100	98	2	-58
Database	demo	NpDb Adm.	58	100	98	200	-40
WEB	S3demo	www.nbc.com	69	80	98	400	-29

Reports:
o   Daily
o   Weekly
o   Monthly

FIGURE 9.7   Service level view.

NextPoint Networks: Weekly Interface Correlation Report

Object: cisco2509 ALTER.NET, Interface, ifindex2/Serial0 (Speed 384 K)

Date: from 11 September 1999 to 17 September 1999
             (Server time zone = Eastern Daylight Saving Time)

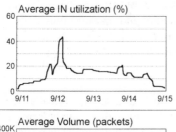
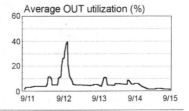
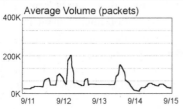
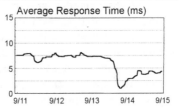

FIGURE 9.8   Network response time overview.

Daily Protocol Distribution and Analysis

NetScout Probe: 22 March, 1999

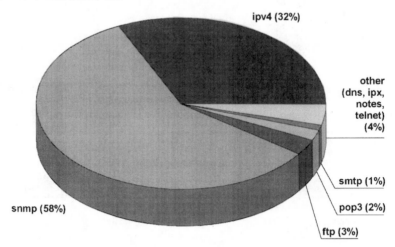

This Pie chart represents 12.6 % of line bandwidth

FIGURE 9.9   Daily protocol distribution and analysis.

basis. The reports also demonstrate when network service expectations as well as important objectives of business-centric management have been met. S3 users can create and access reports tailored to their needs. The Web interface allows this from the convenience of the users' desktop. In addition to the predefined reports, the standard database interface means that any generic reporting package may be used instead.

The NextPoint S3 architecture consists of a robust infrastructure of servers and lightweight intelligent agents to provide scalability and manage complexity. At the same time, the Web interface isolates the user from the complexity with push technology. Figure 9.10 shows the components of the S3 architecture.

The components are:

Distributed server: It provides an engine for real-time and historical analysis, alarm/event handling, and other tasks. The automated task director manages automation routines such as the router configuration management task. It features a modular, object-oriented design for scalability.

Intelligent agents: They provide the user perspective by residing on strategic resources throughout the enterprise — application servers, workstations, network hubs, etc. They are built to be lightweight to avoid burdening production systems with management overhead.

Database: An ODBC-compliant API layer isolates NextPoint S3 system tasks from proprietary databases. S3 works with multiple databases, such as Oracle and MS Access.

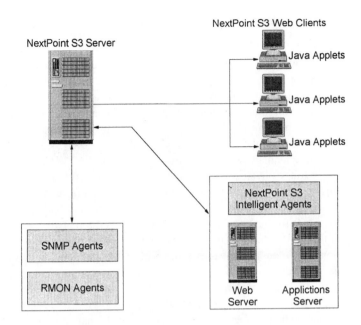

FIGURE 9.10   Architecture of NextPoint S3.

Web interface: The full-function Java interface features push technology that provides dynamic information distribution, flexible access, and ease of administration. User authentication ensures system integrity.

Plug-in modules: Modular options allow users to tailor S3 to specific environments. Frame relay, LAN, and WAN are some of the available options.

In summary, S3 benefits are as follows:

Enhances quality of service
- Tracks impact of network infrastructure on business user (application availability, response time, etc.)
- Complements business user perspective with operational details of network performance
- Measures and demonstrates network service with meaningful reports
- Monitors network service levels against business-centric objectives

Improves operation staff productivity
- Enhances information distribution using fully Web-based push technology
- Automates real-time fault isolation
- Automates routine IP services operations (e.g., address management)
- Tracks enterprise-wide router configuration changes automatically

Reduces network cost-of-ownership
- Optimizes fault detection to reduce down time
- Enables network tuning to minimize costly network slowdown

9.4.3 APPLICATION FLOW MANAGEMENT FROM
NETSCOUT SYSTEMS, INC.

The application flow management solution is an approach to manage enterprise networks by tracking the flow of applications. It connects network management to the needs of the business and establishes a common ground for communications between network operations staff and the users of the network. This family of products gives users the means to collect, analyze, and report the precise information required for network control.

Application flow management addresses all aspects of network traffic, from the applications layer to the physical layer. It instruments the network with the coverage appropriate to its role in the enterprise, then gathers, aggregates, reduces, and forwards the network data when and where it is needed and transforms it into meaningful, usable information. The data is then conveyed to network and business managers to manage the network that runs the business.

Application flow management uses a three-layer architecture. Layer one helps network managers address operational problems by placing probes in strategic locations throughout the network. In Layer two, the NetScout server performs data polling and aggregation, while Layer three allows users to view this data in an easy-to-manage format, whether they are using NetScout Manager Plus for detailed analysis, WebCast to view network data via the Web, or AppScout, which allows users to track certain business-critical applications and their effect on the network.

For enterprise-wide access to network applications traffic, NetScout Systems offers a comprehensive family of data sources:

- NetFlow Monitor and Resource Monitor probe firmware
- VLANs and switched LANs
- Ethernet, fast Ethernet, fast Etherchannel, and token ring probes
- Subrate T1/E1 and T3/E3 WAN probes
- FDDI/CDDI probes
- OC3c ATM probes
- NetScout Remote Agent for Windows NT

NetScout server performs distributed polling and data storage, aggregation, sorting, and reduction of data to keep management traffic down while providing RMON2-based information.

Data presentation is what the network manager sees regarding the network. Timely and actionable network traffic information is provided to all users.

NetScout Manager Plus monitors, analyzes, and reports on data gathered from RMON probes throughout the network. It employs a suite of more than 40 integrated applications and provides information for troubleshooting, managing service level agreements, performance tuning, capacity planning, and enforcing network policies.

AppScout provides a reporting system on the performance and health of mission-critical applications traffic, for line managers who need information on their applications but do not require the level of detail offered in other products. It provides

Response Time History

Recorded: Wednesday, September 15, 1999

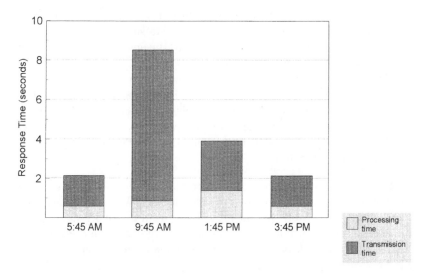

FIGURE 9.11 Response time history by NetScout.

Web-based, RMON2, and ARM MIB data, displaying performance and usage information for specified applications.

AppScout's application flow monitor provides real-time, Web-based information on application performance over an enterprise network. AppScout presents a comprehensive view of real application traffic flows and identifies the application's impact on the network as well as the network's impact on the application. This framework provides a shared perspective from which network and application managers can understand the extent to which applications are leveraging and/or impacting the power of the network.

IT professionals share responsibility for supporting the core of the business — the networked applications and the end users who use them. In many cases, these organizations lack a common management solution from which to work, thereby impeding rapid problem identification and resolution. The AppScout application flow monitor provides IT professionals with a much needed conduit to the information necessary to set realistic expectations and continuously report on how well the network is meeting the needs of its users and the business. When there is a problem, AppScout enables IT professionals to better collaborate, quickly resolve performance issues, and restore services to meet the business' needs. Figure 9.11 shows the response time history for selected applications.

This product is a valuable tool for the ongoing, proactive management of modern, business-critical enterprise networks. AppScout provides the link between application and network performance, offering a shared perspective for

- Avoiding disaster during the rollout of new applications and reoccurring upgrades
- Quickly identifying the source of apparent performance problems (Is it the network or the server?)
- Preventing capital waste by verifying that the network is properly provisioned and operating at sufficient capacity to support prioritized traffic flows; this helps to avoid unnecessary investments in new network hardware or additional WAN bandwidth
- Managing network policies — determining what policies are required and how the policies are working, as well as identifying any violations of the policies; these are critical factors to delivering expected application service levels dictated by policy-based networks
- Enabling the development and measurement of service level management for networked applications

WebCast works with NetScout Manager Plus and the NetScout server to let users access reports and alarms at any time via the WWW. Users can see how activities affect the network and then reschedule them to avoid peak usage times. They can also use the Web to see where additional resources are needed.

NetScout Systems' network monitoring solution addresses four basic network and business management challenges:

1. Fault isolation and troubleshooting — Network outages and slowdowns result in lost productivity, lost revenue, and network management staff burnout. Application flow management lets users fix these problems by measuring application flows, finding bottlenecks in real-time, localizing problems quickly, and solving remote problems with centralized expertise.
2. Policy enforcement — Policies that help the network run more efficiently or more closely aligned to business objectives include limiting access to the Internet for nonbusiness purposes; scheduling runs to better balance traffic loads; setting guidelines for e-mail attachments and downloads; basing network access policies on departmental functions; and initiating charge-backs or report-backs to control usage.
3. Performance tuning and capacity planning — Capacity planning starts with collecting consistent data regarding network link loading, application components, and response time information. By integrating this data, users can measure how much bandwidth is being used by mission-critical applications. This lets users assess specific tactics and watch the immediate effect. The bottom line is that users can plan capacity and tune performance according to the network's business usage by using application flow monitoring instrumentation.
4. Service level management — Service level management helps users connect technology to business goals. The relationship between IT management and network usage can often be challenging. Application flow management delivers service level metrics that users can understand, which in turn supports accountability. This keeps the service providers honest and fosters dialogue about reasonable expectations.

Application flow management concentrates on the traffic carrying business-critical applications. The premise behind this approach to management is that whatever the underlying connectivity, the network's reason to exist is to carry business-critical traffic from desktop to desktop and from data centers to users. Tracking these flows is essential and directly connects resource usage to business goals.

9.4.4 VeriServ from Response Networks

VeriServ is designed for both enterprise IT organizations and application service providers. By measuring the quality of application delivery to the end user directly through active, real-time testing, VeriServ verifies the performance of mission-critical client/server, intranet, and extranet applications such as resource planning, decision support, groupware, team collaboration, and e-commerce. The product is targeted for administrators, operations staff, help desks, and management who need to understand the application service levels of the business units and end customers they support.

With the VeriServ explorer, an operator defines the types of applications to be monitored and the end user locations from which to measure. The operator might also define specific alarm thresholds for response time and throughput. The VeriServ domain controller then instructs the VeriServ agents to begin testing the applications.

The agents execute their tasks and compile statistical summaries, periodically uploading consolidated results to the VeriServ domain controllers. A single VeriServ agent can perform hundreds of real-time transactions to any application that VeriServ supports or to any IP-addressable entity on the network. VeriServ can also execute custom test scripts running between pairs of agents.

When a task exceeds an alert threshold, the VeriServ explorer displays an alarm. Besides providing this real-time "heads up" service quality, the program also summarizes all results and stores them in the VeriServ controller's database for historical reporting. This feature provides managers with the means to guarantee application service levels and to enforce them through verified service level agreements.

The system overhead of a VeriServ installation is very low, unless it is being used deliberately for stress testing. It represents no more strain on the environment than a real user. A typical VeriServ agent installation needs only 48 KB of disk space and uses less than 0.1% of system resources at the workstation.

VeriServ simulates an end user, sending real traffic to actual applications and reporting on the applications' availability and performance. It provides a new class of data not available with traditional management platforms. Reports on performance are available on demand, either from the VeriServ explorer or through a Web browser. Typical reports might include a graphical representation of response time by location, application, organization, and time of day. Other reports could include, for example:

- Select reports by response time, throughput, or both
- Custom report criteria through a Web-based interface
- Reports by percent of goal reached or by absolute performance
- Reports of application performance from remote locations

Critical applications response time and availability

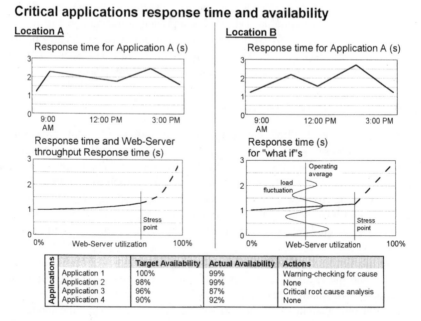

FIGURE 9.12   Report on critical application time and throughput set by VeriServ.

VeriServ reports on "percent of goal achieved", a simple way of viewing service quality against service goals. It delivers critical application response time and throughput data in a variety of useful forms (Figure 9.12).

Users can monitor response times of mission-critical applications continuously, as seen by end users in various locations. This lets users anticipate and avoid end user frustration caused by application problems and other service degradations — whatever the source of the problem. It also helps users establish and enforce service level agreements with business unit managers and external customers. VeriServ is also valuable as a diagnostic tool for application, network, and system troubleshooting. When it detects an emerging service problem, VeriServ will highlight which service components are affected and whether it is a response time or an availability issue. Users can then tap VeriServ's real-time monitoring data to review problems at a high level. It can also help to diagnose problems through historical trends and patterns. It can stress test components or entire systems by simulating end user–generated traffic. It can also be used to evaluate "what if" scenarios involving geographic distribution of users, new applications, new servers, and new network configurations. VeriServ can even reverify the whole application delivery environment after a planned or unplanned environment outage; for example, a network or system upgrade. Users can determine whether the environment will be able to support thousands of concurrent application sessions simply by escalating the appropriate traffic volumes of individual tasks. Network and IT managers can see exactly how their entire system performs, end to end, using the same data that they use to monitor performance in their day-to-day environment. As a service is being restored,

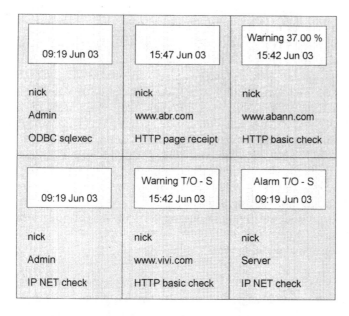

FIGURE 9.13   Overview of critical tasks by VeriServ.

managers can use VeriServ to monitor progress and ensure that the fix has brought the environment back to specified production levels.

The VeriServ operational view shows tasks running from the same VeriServ agent and indicates in color the alarm state of each task:

- Good — Green: currently meeting goals
- Transitional — Yellow or orange: exceeding goals briefly
- Failed — Red: exceeding goals for an extended period

Users and operators can create views simply by clicking and dragging tasks onto the view from a menu. Figure 9.13 displays an overview of critical tasks by VeriServ.

A minimum configuration includes one VeriServ explorer, one controller, and one agent. The actual number of each of the components depends on the requirements of a particular installation. There are no constraints on the maximum number of VeriServ software components. All three can easily run on the same workstation, with little effect on the performance of other applications. Components may also be distributed throughout the network.

VeriServ can be extended to test any client/server application running on a TCP/IP network. The initial release tests the major Internet services, such as HTTP, telnet, FTP, DNS, POP3, SMTP, and NNTP, and any SQL database application through ODBC. For Web-based applications, VeriServ can download a header or a page and can check for the presence of specific text. SQL tests can check and log into an SQL database and execute actual SQL commands. Application modules include major commercial applications such as SAP R/3, PeopleSoft, Oracle, Lotus Notes, Baan, and Microsoft Exchange Server.

All VeriServ components run under Windows NT or Windows 95. The product ships with a Paradox database for smaller installations. For larger implementations, an Microsoft SQL server can be used as a database manager. VeriServ agents run on a certain number of representative client stations within an enterprise. These agents conserve bandwidth between themselves and the VeriServ controller by compressing results data statistically before transmitting it to the database, eliminating the network burden that would result from uploading of unprocessed results.

The VsWeb report engine is installed on a Windows NT machine that controls the database of response time and availability information gathered by VeriServ agents throughout a networked environment. The report engine automatically generates a Web site for viewing reports and automatically updates that Web site each time it runs. In addition, it builds and administers a library of response time and availability reports for viewing and analysis.

Everybody responsible for guaranteeing the performance of network infrastructures can benefit from VeriServ. It helps improve the quality of service (QoS) for internal and external end users in the following ways:

1. Improved network and server response speeds application delivery — VeriServ gives managers the ability to baseline performance and set goals for response time and availability throughout the application delivery environment. It monitors the response of all network devices and servers in real time, helps troubleshoot faults and bottlenecks, and, via its VsWeb report engine, generates Web-based reports automatically for quick viewing and analysis. The result is improved application delivery end to end.
2. Enhanced database and application response improves QoS — VeriServ enables managers to baseline application delivery and database availability and work proactively to improve QoS. It monitors end-to-end performance even in the most extensive environments — including extranets, VPNs, and outsourced infrastructure. VeriServ helps troubleshoot, and in many cases prevent outages, including the intermittent and difficult to diagnose problems with applications. The VsWeb report engine delivers real-time and historical reports on an automatically generated Web site.
3. Higher quality service levels guarantee customer satisfaction — VeriServ is designed to help accomplish some of the most demanding management goals: demonstrating the actual QoS being delivered and documenting response time and availability to verify service level agreements. With this product, IT organizations and service providers can better manage internal and external customer expectations, and the VsWeb report engine makes it simple to deliver detailed reports — showing how well the systems and networking environment are supporting end users.

9.4.5 TRINITY MANAGEMENT SOFTWARE FROM AVESTA

This management application uses intelligent modeling technology to automatically discover and monitor network elements and their relationship. Service level metrics can be measured as well. It then prepares graphical Web-based reports, showing

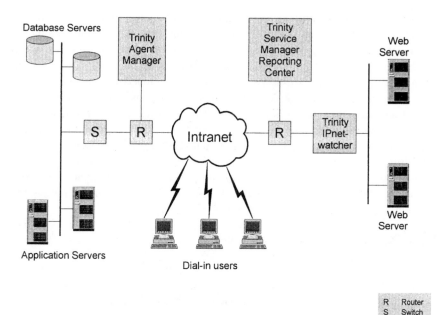

FIGURE 9.14   Architecture of Trinity measurement configuration.

whether revenue-generating applications are meeting service level agreements. Trinity also registers changes made to the network and identifies the root cause of slowdowns, router outages, and other problems generating multiple alarms. That way, network managers get the big picture, instead of large volumes of network management-related data.

Trinity consists of a service manager console and agent managers, both of which run under Unix or Windows NT. Using the service manager's graphical interface, corporate networkers create virtual groups of application and database servers, switches, routers, and clients used by specific applications. For example, they can monitor the devices delivering an order-entry system. They can also copy and paste MIB configurations and thresholds for one or more SNMP agents that monitor the network components. Then, agent managers at key LAN segments or remote sites discover data on group elements, tapping into the agents and relaying critical information to a central Trinity service manager console. Trinity processes the data to determine overall availability (e.g., a weighted average of the availability of the systems, applications, and devices underlying the order-entry system). It then generates graphical reports showing whether service levels are being met. Reports can be customized for network managers, business persons, remote users, and the technically challenged and are accessible via the Web. More agent managers can be added as the enterprise grows. This product helps simplify and speed management. It recognizes the virtual connections between a switch, for instance, and attached servers and clients. If alarms are received from a switch used by the order-entry application, Trinity will report that that application is at risk. Figure 9.14 shows the Trinity configuration.

Trinity also bases reports on data from other management platforms and applications, so companies can make the most of what they already have running. It recognizes topology and event information from Unicenter TNG from Computer Associates, and it can tap into application-monitoring agents from BMC Software via an optional gateway. With another optional package, Ipnetwatcher (Jander, 1999), it can track the performance of Web servers and applications by generating synthetic transactions and measuring results. It also traces virtual LANs built with Cisco routers by recognizing and monitoring the proprietary router information associating logical links with physical ports. It is expected that other vendors' virtual connection capabilities will also be supported.

This product is oriented toward business applications. It can identify problems but cannot solve them. Users still need device-specific element management systems to configure, change, and control devices.

9.4.6 SiteScope from Freshwater Software

The quality of Web sites is a mirror for the company. Public Web sites tell visitors whether the Web site owners appreciate their time and interest. Internal Web sites connected by intranets tell company employees how the Web site owners appreciate their productivity.

Web servers sometimes fail — an event obvious to all types of network and system management tools. Web applications fail for many reasons that are not so obvious. There may be network failures, modem and dialing problems, excessive firewall controls, invalid page links, application bugs, DNS failures, resources overloaded due to a surge in traffic, poor back-end database performance, or other events.

To discover these problems, react to them rapidly, and prevent their recurrence by root-cause analysis, Webmasters need to monitor the Web server, network devices on both sides of the firewall, and application services such as e-mail and databases. Webmasters and site administrators must continuously test the application for response time and accuracy. Discovering that the application does not return the correct answer in a reasonable time and reporting that information to the responsible staff is the first step in improving application availability. However, to decide what actions to take, Webmasters need enough information to pinpoint the cause of the problem. For that reason, they need monitors that can be specialized to supervise Web applications and monitors that can observe from both sides of the firewall. Webmasters want to improve notification by connecting Web site monitoring to traditional systems and network management consoles. In addition, more ambitious Webmasters want a solution that can automatically take corrective actions.

SiteScope is a suite of software tools for monitoring simple and complex e-business Web sites. It tests Web applications by simulating a user's transaction and verifying that the transaction is completed swiftly. SiteScope can be used to monitor multiple servers and devices in single or multiple locations, on both sides of the firewall. It responds to events by sending notification or taking investigative or corrective action. Finally, SiteScope provides management reports on the service level of the Web application. Figure 9.15 shows an oversimplified configuration of SiteScope.

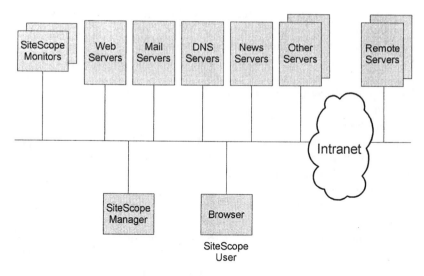

FIGURE 9.15   Configuration of SiteScope.

SiteScope provides the following features (Aldrich, 1999):

- Browser access
- Default configuration
- Comprehensive monitoring
- Customizable monitors and groups
- Automated event response
- SNMP integration with enterprise management
- Support for NT, SGI/IRIX, and Sun/Solaris platforms

SiteScope is a Java server application comprising monitors, the SiteScope framework, a browser-based dashboard, and tools for diagnosis. The browser enables administrators to control and configure SiteScope from any Internet connection. Webmasters use the Web site's security mechanisms to control who can do what with SiteScope monitoring and information. SieScope incorporates the Java Runtime Environment from Sun Microsystems, Java SNMP software from Advent Network Management Group, Java Generic Library from Object Space, Java PerfTools from ORO, and the XML Parser in Java from Microsoft.

The following monitors are available with the product:

URL monitor: It verifies availability and access time for specified URLs to ensure that Web pages are always available in an acceptable time frame. The URL monitor can detect the availability of back-end databases as well as the proper functioning of CGI scripts. String matching capabilities allow users to verify the monitored page's content. Proxy servers are supported to allow Webmasters to see outside their firewall. Leveraging integrated support on the NT platform, SiteScope on NT supports monitoring of secure "HTTPS" URLs.

Disk space monitor: It reports the percentage of disk space currently in use so that Webmasters can act before they run out of disk space. Administrators can customize both warning and error thresholds.

Service monitor: It verifies that specified processes are running, including Web, mail, FTP, news, gopher, and telnet.

Ping monitor: It verifies that specified hosts are available via the network to ensure continuous availability of critical connections (e.g., routers and ISP connections).

CPU monitor: It reports the percentages of CPU currently in use to ensure that administrators and Webmasters know when the CPU is being overloaded.

DNS monitor: It verifies that a DNS server is working and that host names can be resolved.

FTP monitor: It connects to an FTP server and verifies that a file can be retrieved.

News monitor: It connects to a news (NNTP) server and verifies that groups can be retrieved.

Port monitor: It determines whether a service on a port can be connected to.

Memory monitor: This monitor measures virtual memory usage and notifies site administrators before problems occur. Site administrators or Webmasters can set the warning and threshold limits.

Web server monitor: It reads the Web server log and reports data on hits, bytes, errors, hits per minute, and bytes per minute.

Mail monitor: It verifies that the mail server is accepting requests and that messages can be sent and retrieved.

Directory monitor: It checks the directory file count and the total size of all files in the directory to help Webmasters detect unauthorized changes to the Web directory.

SNMP monitor: This monitor returns the value of MIB entries as a response for an SNMP "Get."

Link check monitor: It checks all the internal and external links on a Web site and returns an error if a link cannot be reached.

URL list monitor: This monitor checks a large list of URLs. It is very helpful for ISPs who want to check URLs for their customers.

URL transaction monitor: This monitor verifies that an online transaction can be performed by checking returned page content.

File monitor: It watches the size and age of a file and reports any unauthorized changes. It can even run a script to automatically copy a back-up file to return a file to its proper state if it finds that the file has been changed.

Log file monitor: This monitor can read through an application's log file looking for specific error messages. If it finds an error or other specified message, it can send out notification.

Other monitors: New monitors are added frequently. A continuously updated list of SiteScope monitors is available on the company's home page.

A monitor is a Java-based program. Monitoring can be performed on the Web server to get tighter control of the Web application, outside the Web server, or beyond

Graph of local home page round trip time

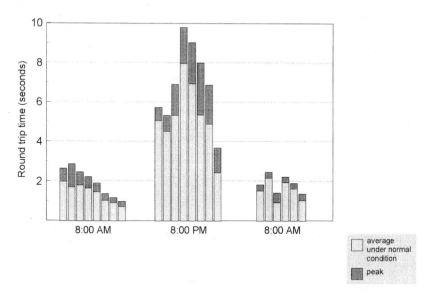

FIGURE 9.16   Graph of local home page by SiteScope.

the firewall to get a clearer view of the user's end-to-end experience. The monitors provide broad, general monitoring, such as Web server health as represented by page retrieval time for a URL. They can also provide very specific monitoring, such as server memory usage or file size and age, useful in tracking down application problems. Administrators can build monitors and scripts for site-specific events using the templates and tools provided with SiteScope.

Figure 9.16 shows a graph of a local home page by SiteScope. Various statistics are highlighted, such as peak, error, warning, and O.K.

With SiteScope, site administrators can:

- Monitor the availability of HTTP and HTTPS URLs and proper function-ing of CGI scripts
- Remotely monitor NT servers from a primary NT server
- Track the status of multiple Web servers from one central panel
- Verify the availability of critical connections (e.g., the ISP's router)
- Watch for low disk space and overloaded CPU
- Ensure that important processes are running
- Generate management reports showing Web server performance
- Forward SNMP traps to an enterprise SNMP platform
- Execute automated recovery scripts

SiteScope helps site administrators keep their sites operational using real-time performance monitoring. When shared resources become constrained or components fail, SiteScope alerts the Web administrator via e-mail or pager in real time. Site-Scope can also initiate automatic error handling by restarting "hung" processes and executing automated recovery scripts.

Immediately after installation, SiteScope automatically begins to collect data by verifying that the Internet/intranet connection is working correctly, checking on CPU and disk space usage, and scanning the Web server log for "hit" information. More than 20 standard monitors and the power to create site-specific custom monitors enables Web administrators to customize the SiteScope monitoring environment. Using Java API, Webmasters can incorporate their own site-specific monitoring scripts into SiteScope, allowing them to retain any homegrown scripts they need for their site and still take advantage of the alert and reporting features built into SiteScope.

9.4.7 ENTERPRISE PRO FROM INTERNATIONAL NETWORK SERVICES

Analysis, reporting, and trending are very useful functions to support performance tuning and capacity planning. However, service level management may need reports delivered in real time. Most of the products introduced here offer this capability. If a company is interested in this kind of reporting but does not want to establish such a service, Enterprise Pro from International Network Services (INS) may be an interesting alternative. It differentiates itself from other products with so called QoS alert capabilities. It immediately notifies network managers when network traffic exceeds a predetermined threshold. And with server monitoring, it helps get to the root cause of problems. Reports can be viewed through e-mail, via Web browsers, or from OpenView consoles, because INS has integrated Enterprise Pro with the OpenView management platform.

Enterprise Pro is installed on a server connected to the LAN as shown in Figure 9.17 (Reardon, 1998). Network managers can set performance thresholds per device, per group of devices, or per segment through a GUI. Using SNMP, the software gathers information from MIBs embedded in routers, switches, bridges, and hubs. It also can collect bandwidth utilization data from RMON1 (remote monitoring) and RMON2 probes throughout the network. As soon as a threshold is exceeded, Enterprise Pro generates a report to the management console or to a browser.

Enterprise Pro can collect data from NT, Novell, and Unix servers. It collects statistics on CPU load, memory utilization, and disk space utilization — information that helps to isolate and resolve problems. Instant updates help corporate networkers start problem resolution before users notice faults or performance degradations. That means they have a better chance of meeting the service level agreements they have signed with various external and internal users.

Enterprise Pro offers information in various levels of detail. Operators, administrators, and network managers can drill down for reports of the latest alarms, results of the last five-minute poll, or a combination of those. It is also possible to customize reports to present information for various sites or groups of devices. Reports can, for instance, be tailored to show activity on the LAN, or on all hubs, or on all routers in a specific geographical area.

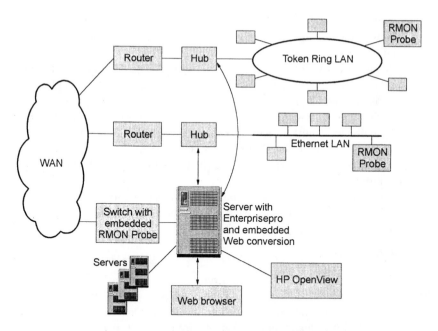

FIGURE 9.17   Enterprise Pro from International Network Services.

Using Enterprise Pro means that certain performance management functions are outsourced to INS. The price/performance ratio is worth considering with this type of outsourcing.

9.4.8   PRONTOWATCH FROM PROACTIVE NETWORKS

ProntoWatch from Proactive Networks may also be used to supervise service level agreements in real time. ProntoWatch goes beyond the capabilities of traditional network monitoring solutions, adding real-time intelligence to the problem detection process, providing proactive notification of trends that could lead to down time, and enabling the user to isolate and eliminate soft faults before they become serious problems. The results are:

- Higher network service levels
- Increased business productivity
- Better competitiveness

This Web-based, turnkey solution is offered at a very affordable price and requires little effort on behalf of the users. Implementation times are very short. Service to the users is delivered electronically over the Web. The benefits of using the ProntoWatch service include:

- Improved service levels — The intelligent early warning support solution leverages the cognitive alarms and 3D telescoping technology to predict and prevent degraded network and application service levels.

- Higher network operations productivity — Real-time intelligent alarms, daily status e-mails, and easy access to network performance and asset information helps streamline network operations.
- Better capacity planning decisions — ProntoWatch monitors and tracks changes in network capacity and performance data so informed planning decisions can be made.
- Savings on equipment and WAN circuit expenses — Infrastructure can be optimized on the basis of complete, dependable, and timely reporting.
- Less resource drain — Due to the adaptive technology and the turnkey approach, management of the underlying technology can be limited to a minimum.
- Favorable payback — Fast installation and user-friendly access to information guarantee a fast payback of investments.

9.5 SUMMARY

Service level management is the highest priority of managing intranets. The state-of-the-art technology alone will not win over users from legacy applications and legacy networks to IP-based intranets. Service and performance are expected to be equivalent to or better than that of legacy environments; at the same time, the use of universal browsers as workstations for accessing applications will significantly simplify operating applications. Financial benefits due to unified training for intranets and Web technology are obvious. The early tools that support service level management concentrate on the measurement of the real response time seen by end users, troubleshooting applications, stress testing applications and networking infrastructures, very flexible reporting, baselining, and optimization. To support all these activities, both emerged data collection technologies and synthetic workloads are used.

10 Trends with Intranet Performance Management

The practical use of Internet technology within the enterprise is changing the way business is conducted. Traditional barriers within the enterprise and between enterprises are pulled down as the interactivity is guaranteed around the clock. Practically, there is no way back to business practices based on personal interaction, paper-based processes, and batch-type data processing.

The dependency of business success on the Internet, intranets, and extranets is continuously growing. The result is that service level expectations are growing as well. Users prefer more and more written service level agreements that also include penalties for noncompliance. Managing intranets is a complex exercise. Managers have to deal with

- multiple suppliers
- systems and networking devices
- high level of service expectations
- new traffic patterns
- unknown user behavior
- emerging new management applications
- emerging new management paradigm

Intranets and extranets are data oriented due to their IP heritage. However, these networks will also expand and embed other communication forms, such as voice and video. IP is standardizing Layer 3 of the communication architecture. Whatever the preferred infrastructure below Layer 3, it must clearly support the rapid growth in IP traffic, as well as legacy services. IP is expected to grow much faster than any other traffic type for the following reasons (Kapoor & Ryan, 1998):

- The number of Internet users continues to double every year.
- New broadband services are offered using xDSL, cable technology, and new networking devices.
- New applications with high bandwidth demand, such as voice and streaming video, are emerging.

- Intranets are standardizing corporate data networking by migrating legacy applications to IP.
- Internet segments are embedded into corporate networks using VPNs and tunneling protocols.
- IP-based applications are usually open and may use different platforms.

The supporting lower layer infrastructure is going to be reassessed. Besides the "traditional" IP-ATM-Sonet/SDH combination, other options are also under consideration. This alternative is conservative, least risky from the engineering point of view, and achievable now. It may, however, lead to low efficiency and high costs. Enhanced frame relay may substitute ATM everywhere, offering lower costs at good quality of service (QoS). However, in certain areas there are no QoS standards available. ATM transport may eliminate the Sonet/SDH layer, offering ATM ring functions similar to distributed ATM switches. There are just a few vendors who consider this option. ATM/IP hybrids on the basis of Sonet/SDH would reduce the number of routers with the result of lower management expenses, This technology is in the test phase, still unproven. IP over Sonet/SDH eliminates the ATM layer completely. If megarouters are at cost parity with ATM switches, this alternative would be the low-cost IP delivery solution. This technology is unproven; operating costs of megarouters are difficult to predict. Optical IP would be the lowest cost delivery of IP services. In this case, IP is directly connected to the optical subnetwork of Layer 1, neither using ATM, nor Sonet/SDH. It is the least proven technology; there are serious concerns about fault management with this alternative.

Intranet management can use experiences from systems and network management. The principal processes, such as fault, configuration, performance, security, and accounting management are the same. Significant extensions are expected with performance and security management functions. Extensions mean the reimplementation of existing pieces combined with emerging management applications and tools. Most of the emerging tools are point solutions. Log file analyzers, traffic monitors, load balancers, and traffic shapers represent the best solutions with little integration. Web server management and related optimization tools can work with management platforms, but this is not always the case. Special tools supporting look-through measurements show more integration by piecing together service metrics from different segments of intranets. Integration is on the way. The management software market is moving rapidly to support Web technologies as the basis for future enterprise management solutions. This approach offers openness and rapid deployment.

Emerging management tools for intranets are still the exception. They are expected to support the basic standards of Web-based management: XML and CIM (Herman & Forbath, 1998). XML is a recently approved standard for structuring Web-based documents and the data they contain. It focuses on the content of documents and complements HTML, which really focuses on how to display documents. XML will allow Web-based information to be much more easily searched and exchanged and will allow Web-based applications to exchange structured information in a standardized way. CIM will support a variety of distributed access methods, including CORBA/IIOP and COM/DCOM from Microsoft. In addition, cross-platform remote access to CIM data is enabled through the newly released XML encoding of CIM.

Additional key technologies that might be needed for this type of integration include directory systems, LDAP, push, secure socket layer, HTML mail, and secure HTTP (Herman & Forbath, 1998).

The big benefit of this approach to tool integration is that it allows each vendor to use its own management agents and data collection and processing solutions as required by its own devices. CIM and XML enable information exchange; browsers offer unified access to information. The dependencies among different management software suppliers are reduced, resulting in easier change management.

Managing intranets and extranets together with Internet technologies may cause significant organizational changes as well. Internet technology has been promoted to higher hierarchical levels. Usually, it is a Vice President level, reporting to the CIO or CTO; in rare cases even to the COO or CEO. Such a promotion demonstrates the importance to businesses. Rapid adoption of Internet standards and services is unifying and simplifying enterprise computing and networking infrastructures and linking it to global information services. In this process, businesses are extending their reach into new markets and working with their business partners more closely. Electronic commerce is restructuring sales channels and support relationships, reducing organizational overhead, and altogether increasing productivity. All these benefits justify the promotion of persons in charge of electronic commerce and Internet technologies.

Outsourcing intranet management is a valid option for top management. However, the golden rule of outsourcing indicates that only well-understood, mature management processes should be outsourced. Intranet management has not yet reached this level of maturity.

Intranet management requires dynamic approaches in an ever changing systems and networking environment. This book helps Webmasters and site administrators with the evaluation of management processes and tools, to select the right mix of tools and the right deployment strategies in the field of intranet performance management.

References

Aber, R.: xDSL supercharges copper, *Data Communications,* March 1997, pp. 99–105.

Aldrich, S.: Freshwater's Web Application Management, Patricia Seybold Group e-Bulletin, January 21, 1999.

Bobrock, C.: Web developers follow old scripts, *Interactive Week,* November 2, 1998, p. 29.

Bock, G. E.: Microsoft Site Server — Organizing and sharing the contents of a corporate intranet, Workgroup Computing Report, Patricia Seybold, August 1998.

Bruno, L.: IP balancing act: Sharing the load across servers, *Data Communications,* February 1999, p. 29.

Gibbs, M.: Pinning down network problems, *Network World,* March 2, 1998, p. 43.

Herman, J., Forbath, T.: Using Internet technology to integrate management tools and information, http://www.cisco.com/warp/public/734/partner/cmc/bmi_wi.htm, 1998.

Hoover, M.: Balancing act, *Network World,* June 14, 1999, p. 49.

Huntington-Lee, J., Terplan, K., Gibson, J.: *HP OpenView — A Manager's Guide,* McGraw-Hill, New York, 1996.

Jander, M.: Clock watchers, *Data Communications,* September 1998, pp. 75–80.

Jander, M.: Network management, *Data Communications,* January 1999, p. 75.

Kapoor, A., Ryan, J.: Reassessing networks for an IP architecture, *Telecommunications,* October 1998, p. 48.

Larsen, A. K.: All eyes on IP traffic, *Data Communications,* March 1997.

Leinwand, A., Fang, K.: *Network Management — A Practical Perspective,* Addison-Wesley, New York, 1993.

Mason, R. P.: WebSpective for Web Operations Management, White Paper, International, Data Corporation, Framingham, MA, 1999.

Powell, T.: An XML primer, *InternetWeek,* November 24, 1997, pp. 47–49.

Reardon, M.: Global load balancers — pointing web requests in the right direction, *Data Communications,* August 1999, pp. 61–72.

Reardon, M.: Traffic shapers: IP in cruise control, *Data Communications,* September 1998, p. 67.

Rubinson,T., Terplan, K.: *Network Design — Management and Technical Perspectives,* CRC Press, Boca Raton, FL, 1998.

Santalesa, R.: Weaving the Web fantastic — Authoring tools, *InternetWeek,* November 17, 1997.

Schultz, K.: Two tools for monitoring your Web site, *InternetWeek,* October 27, 1997, pp. 60–61.

Sturm, R.: *Working with Unicenter TNG,* QUE Publishing, Indianapolis, IN, 1998.

Taylor, K.: Internet access: Getting the whole picture, *Data Communications,* March 1996, pp. 50–52.

Terplan, K.: *Effective Management of Local Area Networks, Second Edition,* McGraw-Hill, New York, 1996.

Terplan, K.: *Telecom Operations Management Solutions with NetExpert,* CRC Press, Boca Raton, FL, 1998a.

Terplan, K.: Web-based systems and network management, Xephon Briefing, London, October 14, 1998b.

Terplan, K.: *Web-Based Systems and Network Management,* CRC Press, Boca Raton, FL, 1999.

Terplan, K., Huntington-Lee, J.: *Distributed Systems and Network Management,* Van Nostrand Reinhold, New York, 1995.

Acronyms

3D	Three dimensional
4GL	Fourth-generation language
ACL	Access control list
API	Application programming interface
ARM	Application response measurement
ARP	Address resolution protocol
ASIC	Application-specific integrated circuit
ASP	Active server pages
AUO	Active user object
BGP	Border gateway protocol
BSS	Business support system
BUI	Browser user interface
CBQ	Class-based queuing
CDF	Channel definition format
CGI	Common gateway interface
CIM	Common information model
CIR	Committed information rate
CLEC	Competitive local exchange carrier
CLI	Command line interface
CMIP	Common management information protocol
CMISE	Common management information service element
CNM	Customer network management
COM	Common object model
CORBA	Common object resource broker application
COS	Class of service
CPE	Customer premises equipment
CRC	Cyclic redundancy check
CSS	Cascading style sheet
CSV	Comma separated value
DATM	Distributed application transaction measurement
DCE	Distributed computing environment
DCOM	Distributed common object model
DDM	Distributed database manager
DEN	Directory-enabled networks
DES	Data encryption standard
DHCP	Dynamic host configuration protocol
DHTML	Dynamic hypertext markup language

DLL	Dynamic link library
DME	Distributed management environment
DMI	Desktop management interface
DMTF	Desktop Management Task Force
DNS	Domain name service
DOM	Document object model
DSI	Data source integration
DSSSL	Document style and semantics specification language
DTD	Document type definition
EC	Electronic commerce
ELEC	Enterprise local exchange carrier
FDDI	Fiber distributed data interface
FTP	File transfer protocol
GUI	Graphical user interface
HMMP	Hypermedia management protocol
HMMS	Hypermedia management schema
HMOM	Hypermedia object manager
HTML	Hypertext markup language
HTTP	Hypertext transfer protocol
HTTPS	Hypertext transfer protocol secure
ICMP	Internet control message protocol
IFS	Installable file system
IIS	Internet information server
IKE	Internet key exchange
ILEC	Incumbent local exchange carrier
IMT	Inductive modeling technology
IP	Internet protocol
ISAM	Index sequential access method
ISP	Internet service provider
ISV	Independent software vendor
ITA	IT/Administration
ITO	IT/Operations
JDBC	Java database connectivity
JEL	Java event list
JMC	Java management console
JMI	Java management interface
LAN	Local area network
LDAP	Lightweight directory access protocol

MA	Management application
MAC	Message authentication code
	Media access unit
MD 5	Message digest 5
MDC	Modification detection code
MIB	Management information base
MIF	Management infomation format
MO	Managed object
MOF	Managed object format
MRP	Message routing policy
MRT	Mini router
MSS	Microsoft SQL server
	Marketing support system
NCSA	National Center for Supercomputing Applications
NFS	Network file system
NNM	Network node manager
ODBC	Open database connectivity
OEM	Original equipment manufacturer
OM	Object manager
OSD	Open software description
OSDFP	Open shortest path first
OSS	Operations support system
PDU	Protocol data unit
POP	Point of presence
RBOC	Regional Bell Operating Company
RDBMS	Relational database management system
RDF	Resource definition framework
RIP	Routing information protocol
RMI	Remote method invocation
RMON	Remote monitoring
ROI	Return on investment
RPC	Remote procedure call
RSVP	Resource reservation protocol
SDN	Software-defined network
SGML	Standardized generalized markup language
SHA-1	Secure hash algorithm
SLA	Service level agreement
SMI	Structured management information
SMIL	Synchronized multimedia integration language
SMS	Systems management server

SMTP	Simple mail transfer protocol
SNA	Systems network architecture
SNMP	Simple network management protocol
SQL	Structured query language
SRM	Site resource management
SRT	System route table
SSL	Secure socket layer
TCP	Transmission control protocol
TME	Tivoli management environment
TMN	Telecommunications management network
TNG	The next generation
UDP	User datagram protocol
UML	Unified object model
UNC	Universal naming convention
URC	Universal resource citation
URL	Universal resource locator
URN	Universal resource name
VAN	Value-added network
VPN	Virtual private network
WAN	Wide area network
WBEM	Web-based enterprise management
WDM	Windows driver model
WMI	Windows management instrumentation
WMO	Web management option
WWW	World Wide Web
XLL	Extensible linking language
XML	Extensible markup language
XSL	Extensible style language

index